I0057002

A WALK THROUGH GEOLOGIC TIME FROM MT. BAKER TO BELLINGHAM BAY

GEOLOGIC SCULPTING OF THE WHATCOM COUNTY LANDSCAPE

Don J. Easterbrook

Copyright © 2010 by

Don J. Easterbrook

Bellingham, Washington

All rights reserved. No part of this book may be reproduced in any form
or by electronics without permission in writing from the publisher.

ISBN 978-0-9842389-6-5.

Library of Congress Control Number: 2010937363

Published by Chuckanut Editions, Bellingham, WA. in 2010

Printed by Village Books

Bellingham, Washington

Front cover: Mt. Shuksan, Picture Lake at Heather Meadows. Photo by D.J. Easterbrook
Back cover: Mt. Baker. Photo by D.J. Easterbrook

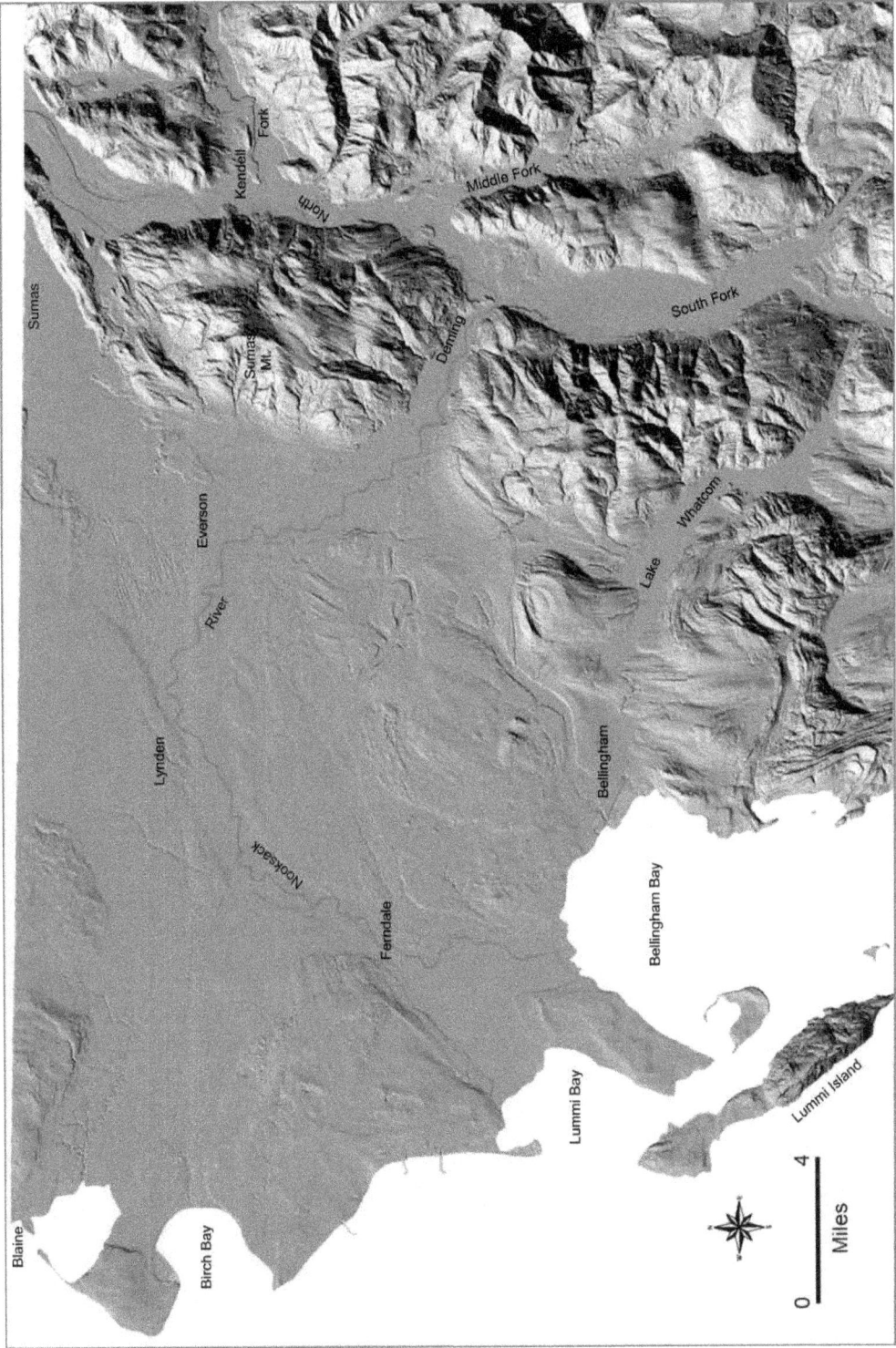

Blaine

Birch Bay

Sumas

Kendall Fork

North

Middle Fork

Deming

Sumas Mt.

South Fork

Lynden

Everson

River

Nooksack

Whatcom

Lake

Ferndale

Bellingham

Lummi Bay

Bellingham Bay

Lummi Island

0

4

Miles

3

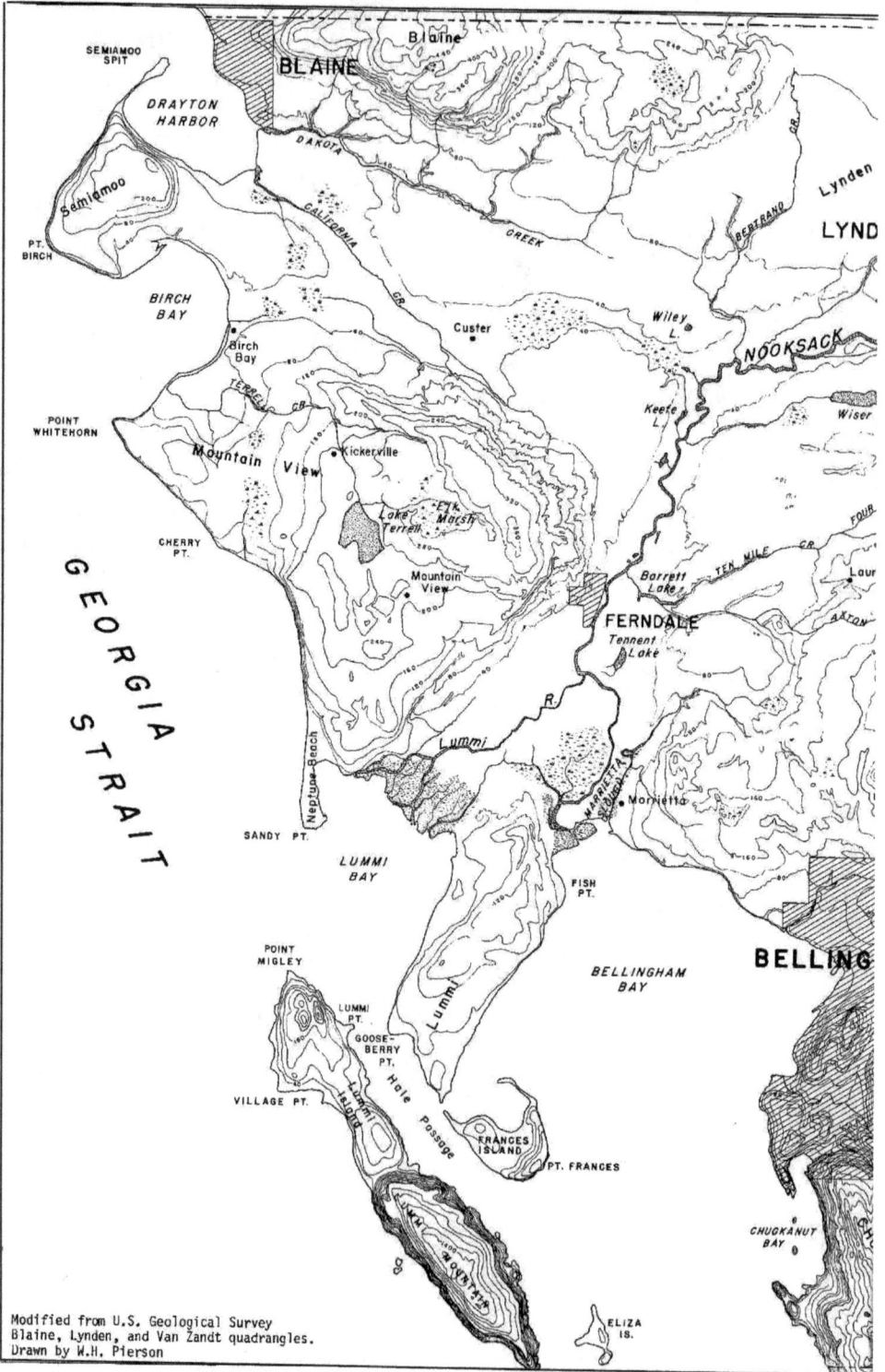

Modified from U.S. Geological Survey
Blaine, Lynden, and Van Zandt quadrangles.
Drawn by W.H. Pierson

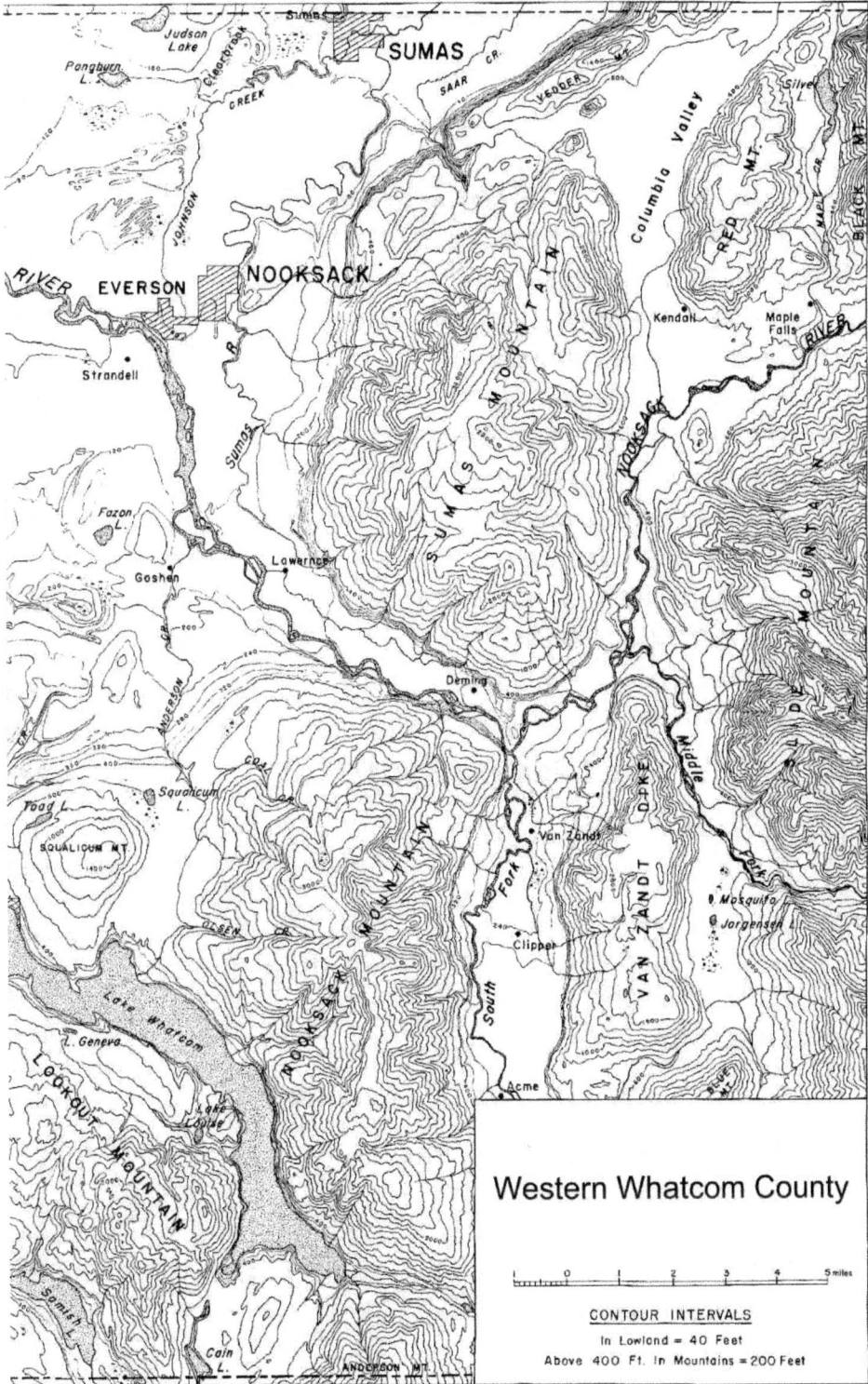

SUMAS

NOOKSACK

EVERSON

RIVER

Judson Lake

Pangburn L.

Clearbrook

CREEK

JOHNSON

SAAR

Sumas CR.

VEDDER MT.

Columbia Valley

RED MT.

Silver

BLACK MT.

MAPLE CR.

Kendall

Maple Falls

RIVER

Strandell

Sumas R.

Fazon L.

Goshen

Lawrence

S U M A S   M O U N T A I N

NOOKSACK CR.

MIDDLE   MOUNTAIN

Deming

Squalicum L.

Toad L.

SQUALICUM MT.

1400

ANDERSON CR.

OLSEN CR.

N O O K S A C K   M O U N T A I N

South Fork

Fork

Van Zandt

VAN ZANDT DIKE

Middle

Clipper

Mosquito

Jorgensen L.

Acme

BLUE MT.

Lake Whatcom

L. Geneva

LOOKOUT MOUNTAIN

Lake Louise

Somish R.

Cain L.

ANDERSON MT.

Western Whatcom County

1   0   1   2   3   4   5 miles

CONTOUR INTERVALS

In Lowland = 40 Feet

Above 400 Ft. In Mountains = 200 Feet

5

CONTENTS

GEOLOGIC TIME SCALE AND GEOLOGIC EVENTS IN WHATCOM COUNTY.

ERA	PERIOD	AGE	EPOCH	SOME EVENTS IN WHATCOM COUNTY
CENOZOIC (Age of Mammals)	QUATERNARY	10,000	HOLOCENE	DEPOSITION OF DELTAS, FANS, SPITS, AND FLOODPLAINS
			PLEISTOCENE ICE AGE	ERUPTIONS OF MT BAKER; GLACIATION BY CONTINENTAL ICE SHEETS; SUBMERGENCE OF THE LOWLAND; ANCIENT SHORELINES TO 600'; BUILDING OF MANY GLACIAL MORAINES AND DEPOSITION OF OUTWASH PLAINS
	TERTIARY	2 million / 50 million / 65 million	PLIOCENE MIOCENE OLIGOCENE EOCENE PALEOCENE	UPLIFT OF CASCADE MTS / CHUCKANUT FM
MESOZOIC (Age of Reptiles)	CRETACEOUS JURASSIC TRIASSIC	250 million		NOOKSACK GROUP / CULTUS FM.
PALEOZOIC (Age of invertebrates)	PERMIAN PENNSYLVANIAN MISSISSIPIAN DEVONIAN SILURIAN ORDOVICIAN CAMBRIAN	400 million / 600 million		CHILLIWACK GROUP / YELLOW ASTER COMPLEX
	PRECAMBRIAN	4.5 billion		

INTRODUCTION

My roots run deep in Whatcom County, and although I now live in a city, my soul resides in the mountains! My great grandparents were Whatcom County pioneers, beginning with a homestead near Kendall, then moving to Sumas where my dad was born in a log cabin on the Easterbrook Road. I was also born at home on the Easterbrook Road (but not in the log cabin).

My parents loved the mountains, and from the time I was about three years old, we took frequent family hikes in the North Cascades. I grew up roaming the high country and the lowlands of Whatcom County. My brother, Bill, and I spent most of our summers hiking and fishing high mountain lakes. Since then, I've seen a good part of the world as a geologist and have come to realize that Whatcom County is not only a scenic wonder, but contains a greater variety of geologic features than any comparable area in the world. Ours is a truly unique region.

Whatcom County is not only one of the most scenic areas in the world, but is also one of the most interesting geologically. From the ocean it is but an hour and a half drive to the wild beauty of alpine glaciers and the varied geologic features of the North Cascades. Recorded in the rocks of the region is a tale as enthralling as the modern geologic processes that are continuing to sculpt the land.

Organization of chapters

The geologic features described in the book are arranged roughly from oldest to youngest. An attempt has been made to focus on geologic features that can be seen easily in the field. Field guides and maps to specific localities are included in "do it yourself" field trips to see many of the geologic wonders of the county. A glossary of geologic terms is included in the back of the book for reference.

ACKNOWLEDGEMENTS

I would like to thank the many people that helped make this book possible. Ellen and Karen Easterbrook contributed endless hours of editing, proof–reading, and suggestions, as well as Nan and Tom Thomas, Mary and Tim Irvin, Carol and Les Shanahan, Bill Easterbrook, and Lindsey McGuirk. George Mustoe of Western Washington University is a bottomless well of information about fossils in Whatcom County and generously contributed a number of photographs.

CHAPTER 1
THE NORTH CASCADE RANGE

To anyone who loves mountains, the North Cascades are a never–ending enchantment, displaying a broad spectrum of mountain structures, landforms, micro-climates, and scenic beauty. The mountains rise thousands of feet through steep timbered ridges, separated by the V–shaped canyons eroded by energetic streams vigorously dissecting the terrain, to high, glacier–clad peaks. In the North Cascades, resistant crystalline rocks that originated deep within the Earth's crust have risen so recently and forcefully that erosion has produced breathtaking sculpture. Intense incision by streams and glaciers has exposed the deep–seated rocks that now soar skyward as ragged snow-covered peaks and precipices with snowfields, lakes, and glaciers strung with countless waterfalls for which these mountains are named.

Figure 1. The North Cascades. (Photo by D.A. Rahm)

YELLOW ASTER CRYSTALLINE ROCKS—A 1.8 BILLION YEAR OLD PIECE OF THE EARTH'S CRUST

The oldest rocks in the North Cascade Range are the Yellow Aster crystalline basement rocks that represent an ancient piece of the Earth's crust. These rocks were formed far below the surface and pushed up along large thrust faults where they have been exposed by erosion of the overlying rocks. They formed in the earliest part of the Earth's history, the Precambrian Era, about 1.8 billion years ago, and are among the oldest rocks in North America. The Precambrian Era spans the time of formation of the Earth (about 4.5 billion years ago) to the first appearance of abundant fossils about 540 million years ago. Precambrian rocks typically occur in the core of mountain ranges that have been pushed up by tectonic forces and the overlying rocks have been eroded away. These rocks commonly consist of granite gneiss formed by intense heat and pressure far below the Earth's surface.

Figure 2. Ancient Yellow Aster granite gneiss. (Photo by George Mustoe)

11

The Yellow Aster unit includes several different kinds of rock that have been subjected to strong heat and pressure over a long period of time. The rocks consist of (1) granitic gneiss, a rock formed by crystallization of older rocks by intense heat and pressure deep beneath the Earth's surface, (2) intrusive molten granite injected into the rocks, and (3) ancient sedimentary rocks that were later crystallized by high temperatures and strong pressure at great depth below the surface.

The Precambrian rocks at Yellow Aster Meadows (Fig. 3) occur in a four-square mile slab of granite gneiss that has been squeezed up along low angle thrust faults and inter-sliced with other rocks as though shuffled in a giant deck of crustal cards. Isotope dating indicates that they contain minerals that are 1.8 billion years old.

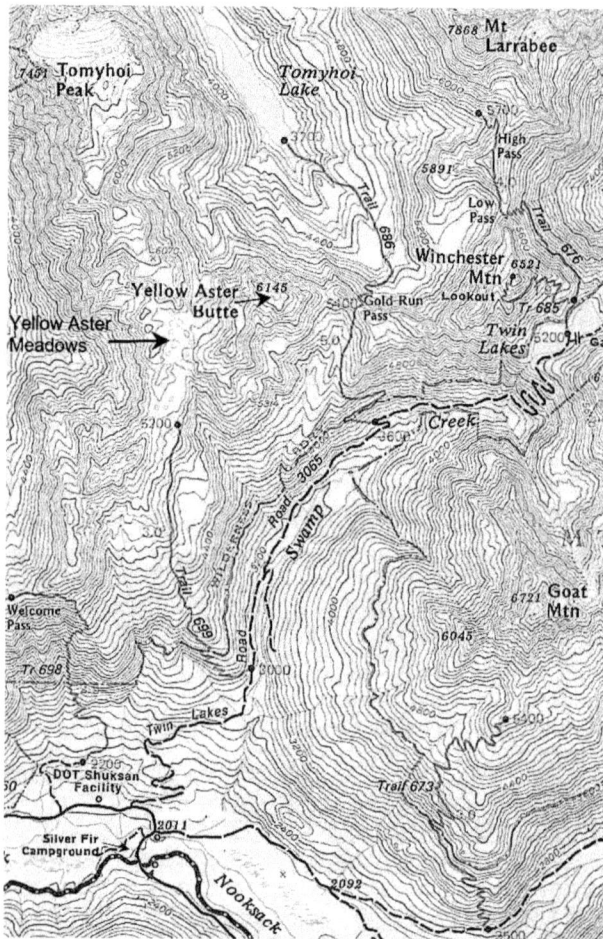

Figure 3. Map of Yellow Aster Meadow north of the Nooksack River and Mt. Baker Highway, west of Twin Lakes. (USGS topographic map.

**Figure 4. Yellow Aster granite gneiss at Yellow Aster Meadow.
(Photo by George Mustoe)**

The best places to see examples of the ancient Yellow Aster rocks are in Yellow Aster Meadow (Figs. 3, 4) and at Schreibers Meadow on the south flank of Mt. Baker near Park Butte (Fig. 5). Yellow Aster Meadow is just south of Tomyhoi Peak and west of Twin Lakes (Fig. 3). It may be reached by U.S. Forest Service trail 699 off the Twin Lakes road north of the Mt. Baker highway about 10 miles east of the town of Glacier.

At Yellow Aster Meadow, bare rock is exposed surrounding small ponds (Fig. 4). Yellow Aster Butte, which rises above the meadow, consists of chert (a marine rock composed of silica) and shale of the Bell Pass Melange (about 225 million years old) (Fig. 6). These rocks are overlain by Darrington phyllite, a former shale that was slightly crystallized by low heat and pressure about 160 million years ago. Both of these overlying rock units are slices of rock that have been thrust over other rocks along low angle faults (thrust faults).

The Yellow Aster rocks occur elsewhere in the North Cascades in small slices and slabs bounded by faults. They can be seen on Park Butte above Schreibers Meadow by hiking cross-country from the Park Butte trail through the forest to the base of a talus slope of loose rock on Survey Point. Large boulders of Yellow Aster rocks at the base of the loose talus consist of granitic gneiss, dated at 1.8 billion years, interlayered with folded marble and cut by dikes (small intrusions of molten rock along fractures) (Fig. 5).

13

Figure 5. Yellow Aster granitic gneiss, dated at 1.8 billion years, and folded marble cut by dikes at Schreibers Meadow. (Photo by E.H. Brown)

Figure. 6. Geologic map of Yellow Aster Meadow, made up of Precambrian Yellow Aster granitic gneiss (dark area). Yellow Aster Butte consists of Darrington phyllite overlying the Yellow Aster rocks along a low angle thrust fault (lines with dark triangles). (Modified from Brown et al., 1987)

15

THE CHILLIWACK GROUP—A 400 MILLION YEAR OLD SEA AND LARGE VOLCANIC ERUPTIONS

400 million years ago, the eastern part of Whatcom County in the area of the present North Cascades lay beneath the sea. Sand, mud, silica, and lime were deposited on the seafloor and later turned into sandstone, shale, chert, and limestone by cementation and compaction. Volcanoes erupted large volumes of lava. Marine organisms living in the sea were buried in the sediments and now occur in the rocks as fossils.

The Chilliwack Group consists of two units, one dominated by volcanic rocks and the other composed largely of marine sandstone, shale, and small lens–shaped beds of limestone. Both the volcanic and sedimentary rocks are thought to have accumulated in an ancient volcanic arc, perhaps similar to the present-day Aleutian arc in Alaska. Most of the volcanic rocks have been slightly metamorphosed under relatively low heat and pressure and the original minerals in the rocks have been altered to green minerals (actinolite and epidote). These minerals give the rocks a characteristic green color so they are known as 'greenstones.'

Most of the limestone is gray, dense, and finely crystalline, formed during the Paleozoic Era (Age of Invertebrates). (see geologic time scale, p. 8). Fossils are rare in the Chilliwack sandstone and shale, but are abundant in small lenses of interbedded limestone that occur on Red Mt., Black Mt., and Sumas Mt. near Kendall.

Marine fossils in the limestone include corals, ostracods (tiny crustaceans known as seed shrimp), crinoids (sea lilies), bryozoa (tiny marine organisms that form colonies), stromatoporoids (a type of reef-forming sponge), and brachiopods (mollusks resembling clams). Much of the limestone deposited during the Pennsylvanian Period (p. 8) (about 300 million years old) of the Paleozoic Era consists of fragments of crinoid stems, a type of sea lily. Some beds consist almost entirely of crinoid stems. Fusilinids, tiny, microscopic single celled foraminifera, are common in the limestone deposited during the Permian Period of the Paleozoic (p. 8).

Coral reefs are common in limestone deposited during the Devonian Period (p. 8), whereas thick lens–shaped beds of limestone are common in the Pennsylvanian and Permian (p. 8) deposits. All of the limestone is interbedded with shale, chert, sandstone, and volcanic rocks.

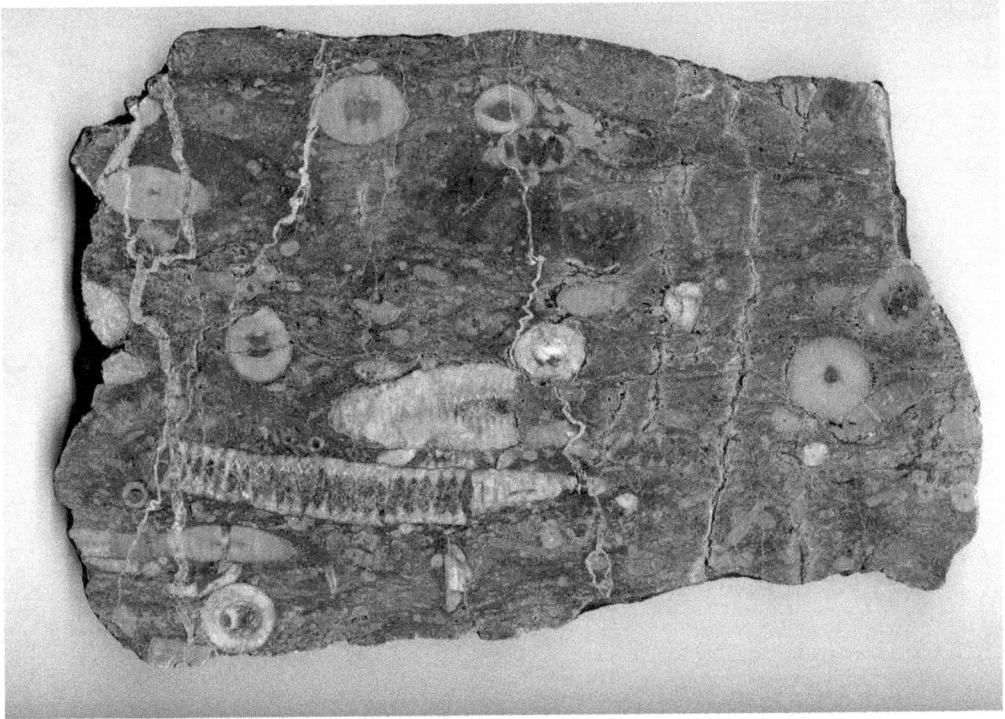

Figure 7. Marine fossils in limestone of the Chilliwack Group. The long columnar fossils are stems of crinoids, a type of sea lily. The circular forms are cross sections of the crinoid stems. (Photo by George Mustoe)

The limestone on Red Mt. (Figs. 8) is about 300-400 feet thick and rests on shale, sandstone, and chert. It is overlain by shale and thick volcanic rocks (Fig. 9). The limestone on Black Mt. is overlain by thick conglomerate composed mostly of cobbles of volcanic rocks. The best places to see the limestone and collect fossils are in the quarries on Red Mt. (Fig. 10).

Figure 8 is a geologic map of Red Mt., Black Mt. and Sumas Mt. showing the distribution of rocks of the Chilliwack Group. The stippled pattern on Red Mt. and Black Mt. represents the part of the Chilliwack Group dominated by volcanic rocks and the lighter area to the north represents the part of the Chilliwack dominated by marine sedimentary rocks. The limestone lenses that have been quarried occur in the Chilliwack sedimentary rocks.

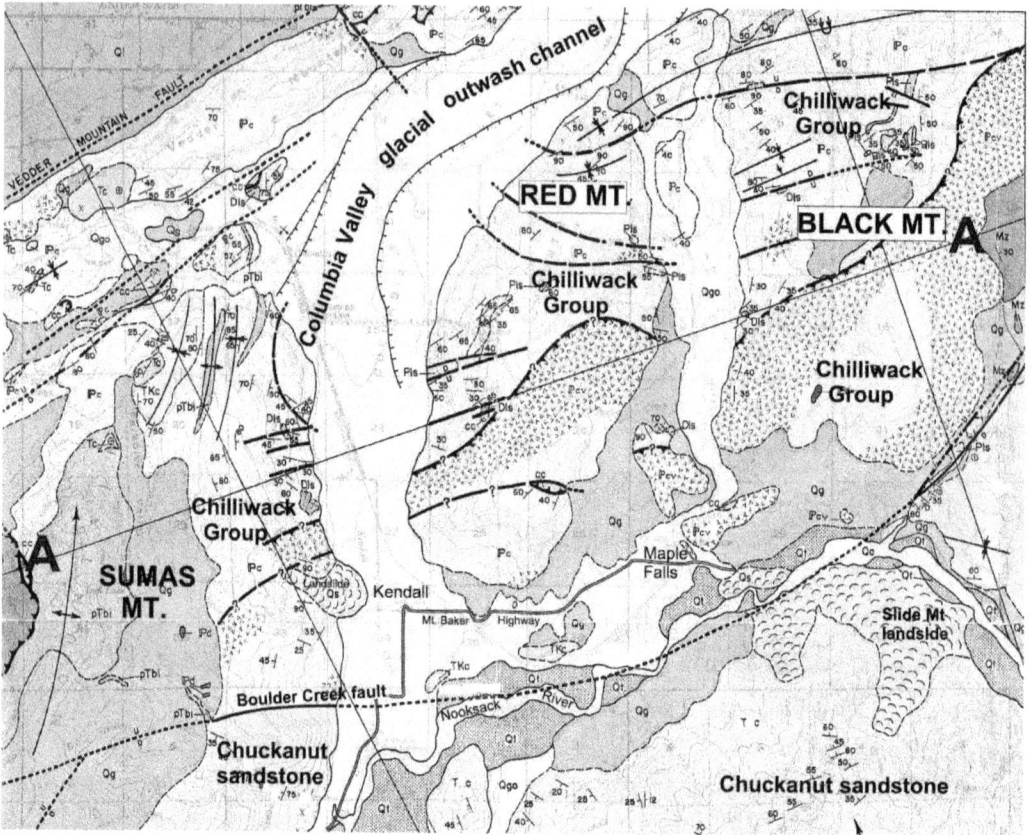

Figure 8. Geologic map of the area north of the Nooksack North Fork showing the distribution of Chilliwack Group rocks and the Chuckanut Formation. (Modified from Moen, 1962)

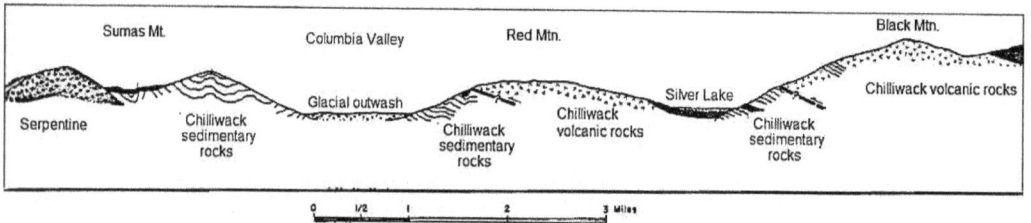

Figure 9. Geologic cross section along the line A-A on the geologic map above. (Modified from Moen, 1962)

A geologic cross section (side view) along the line A-A on the map is shown in Figure 9. On the right hand side of the figure, Chilliwack volcanic rocks rest on top of the sedimentary beds on both Red Mt. and Black Mt., separated by low-angle faults.

18

Limestone quarries

Limestone has been quarried extensively in Whatcom County since the early 1900s, primarily for use in making cement but it is also used for rip-rap because of its high durability. An estimated 10 million tons of limestone were quarried from 1912 to 1965.

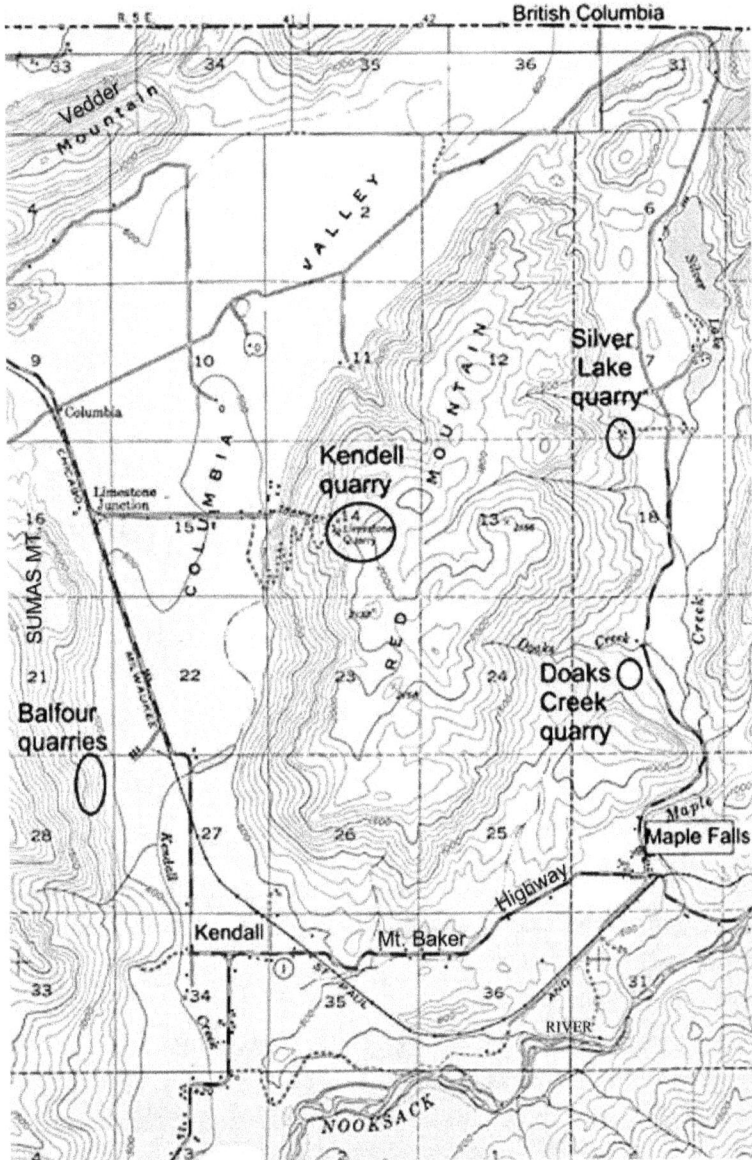

Figure 10. Topographic map of Red Mt. showing the locations of limestone quarries. (USGS topographic map)

The Kendall quarry is on the west side of Red Mt. about 2 1/2 miles north of Kendall (Fig. 10) and is the largest and only active quarry. The limestone there occurs as a lens–shaped bed about 500 feet thick, 4,000 feet long, and 120-700 feet wide. It consists of early Pennsylvanian (about 300 million years old) dark gray, crystalline limestone composed almost entirely of fossil crinoid (sea lily) columns (Fig. 7). In places, the limestone contains a lot of silica and locally passes into chert. The limestone beds trend NE-SW and are tilted 45° to 55° to the southwest. Overlying the limestone are beds of dark gray shale and poorly sorted sandstone with interbedded volcanic rocks. The rocks have been intensely fractured, many of which have been filled with calcite that was precipitated from water passing through fractures in the rock.

The Silver Lake quarry is about 2 1/2 miles north of Maple Falls just west of the south end of Silver Lake (Fig. 10). Crystalline, early Pennsylvanian limestone there trends E-W and is tilted about 45° to the south. It is intensely contorted and sheared and highly fractured with many white calcite veins.

The Doaks Creek quarry is about 1 1/4 miles north of Maple Falls on the road to Silver Lake (Fig. 10). Dark gray, finely crystalline limestone there trends N60°E and is tilted over a range 30° to 70° both to the north and to the south. Beds of resistant Devonian (about 350 million years old) coral several feet thick are interlayered with less resistant non-fossiliferous limestone, giving the rocks a banded appearance. The quarry has been abandoned for many years.

The Balfour quarry is about 1 1/4 miles north of Kendall on the eastern flank of Sumas Mt. (Fig. 10). Dark gray, crystalline limestone there contains Devonian corals and is highly contorted and fractured with many calcite veins. It trends generally E-W and is tilted 50° to 80° to the south. The quarry operated from 1913 to 1929 and then was abandoned when operations were shifted to the Kendall quarry.

Upper Nooksack North Fork

Chilliwack marine sedimentary rocks and volcanic rocks also occur farther upstream in the Nooksack drainage between the town of Glacier and Ruth Creek where they have been thrust faulted over the younger Nooksack marine sedimentary rocks and uplifted in a broad arch whose axis is north–south, parallel to Wells Creek (Figs. 13, 14). A good example of this relationship can be seen at Church Mt. (Figs. 11, 12), which may be observed driving east from Glacier toward Douglas fir campground on the Mt. Baker highway. The upper part of the mountain consists of Chilliwack volcanic rocks, which have been pushed over the younger marine sedimentary rocks of the Nooksack Group along a low-angle thrust fault that crosses the mountain face from the right to Douglas fir campground on the left.

Figure 11. Church Mt. from Mt. Baker Highway below Douglas fir Campground.

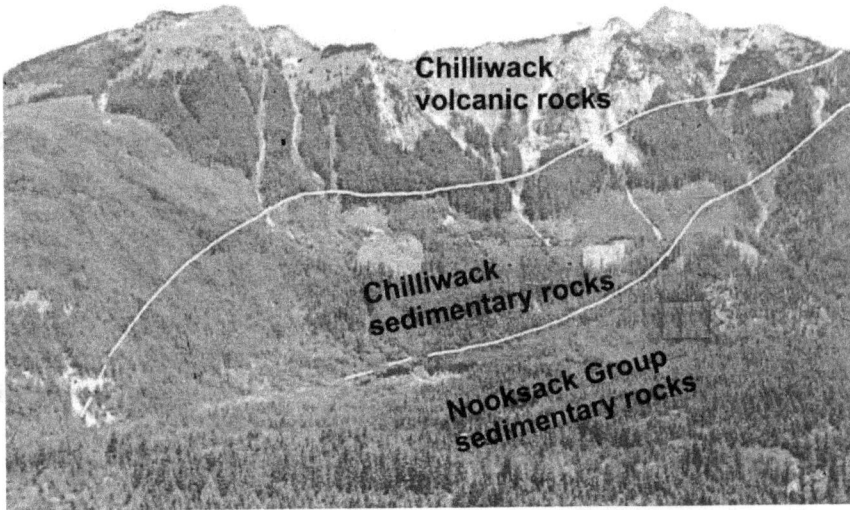

Figure 12. Church Mt. Chilliwack volcanic and sedimentary rocks thrust over the younger Nooksack marine sedimentary rocks.

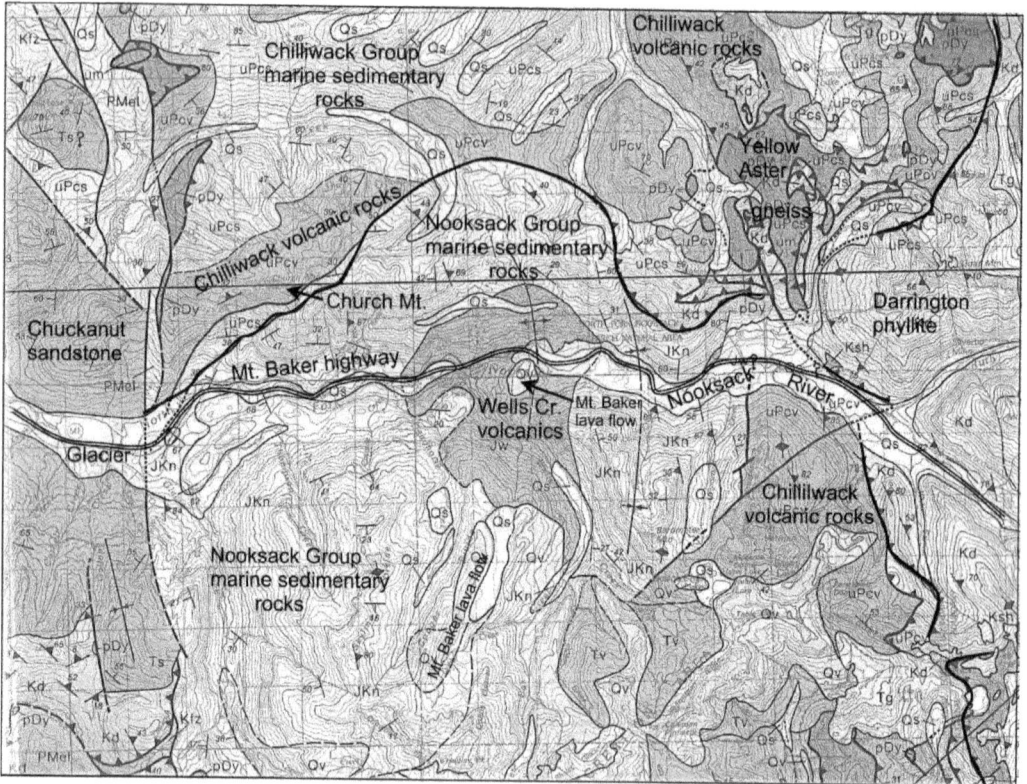

Figure 13. Geologic map of the North Cascades between Glacier and Ruth Creek. (Modified from Brown et al., 1987)

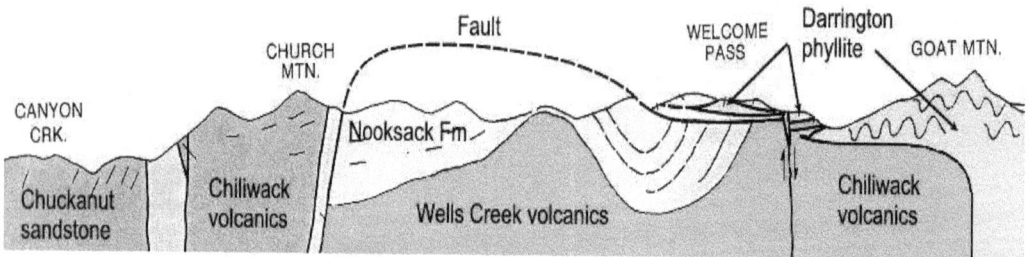

Figure 14. Geologic cross section from Church Mt. to Goat Mt. parallel to the Mt. Baker highway. Church Mt. consists of Chilliwack volcanics that have been faulted over the Nooksack marine sedimentary rocks. The Wells Creek volcanics are exposed by erosion of the top of an anticlinal arch. At Welcome Pass several low angle thrust faults have shuffled Darrington phyllite, Chilliwack sedimentary rocks, and Yellow Aster crystalline rocks together into a stack that overlies Nooksack marine sedimentary rocks. Goat Mt. consists of Darrington phyllite thrust faulted over Chilliwack volcanic rocks. (Modified from Brown et al., 1987)

East of Church Mt., several peaks between the Nooksack River and the Canadian border consist of rocks of the Chilliwack Group. American Border Peak, Mt. Larrabee, Winchester Mt., Goat Mt., and Tomyhoi Peak are all composed of Chilliwack volcanic and sedimentary rocks (Figs. 13–21).

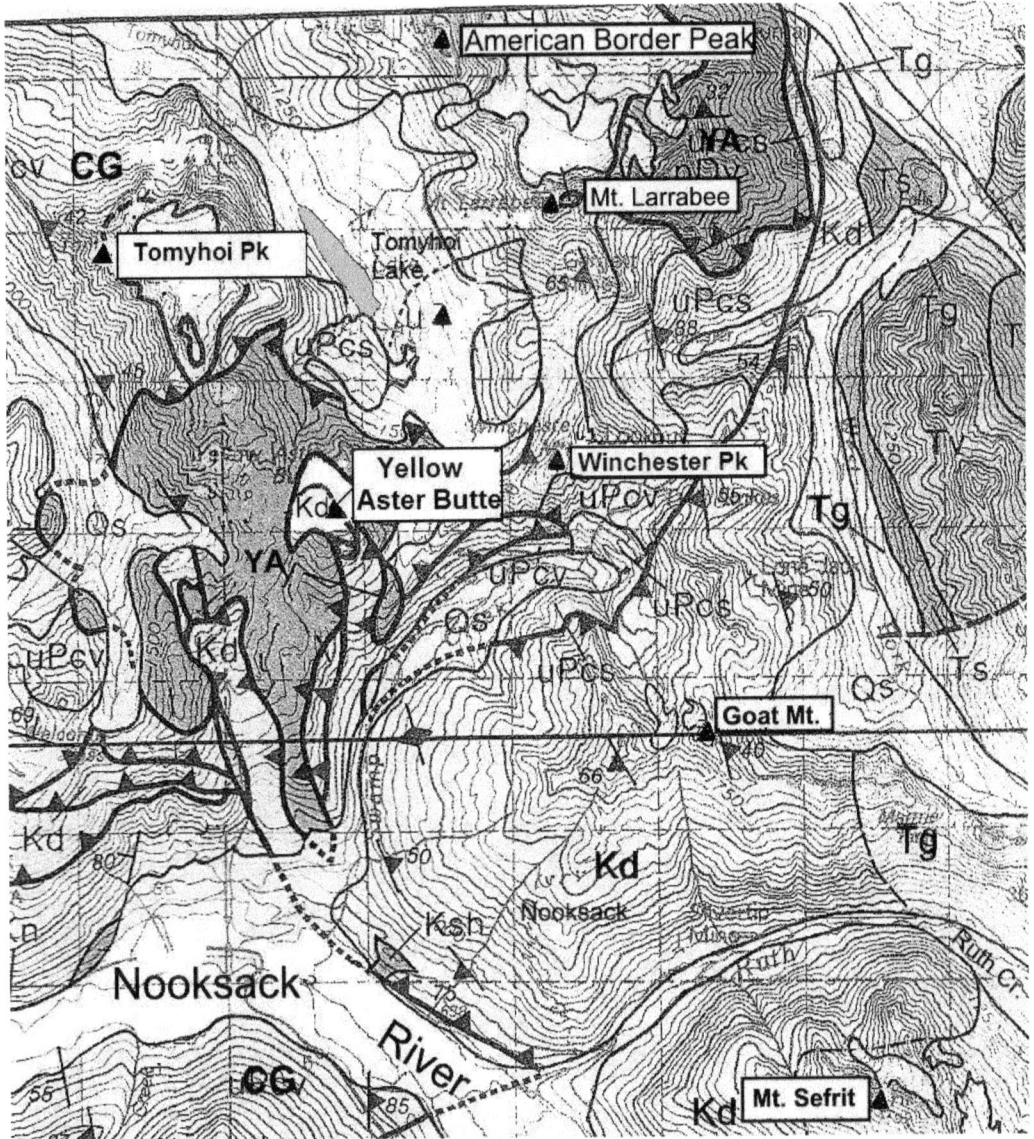

Figure 15. Peaks between the Nooksack River and the Canadian border composed of rocks of the Chilliwack Group. CG = Chilliwack volcanic and marine sedimentary rocks; YA = Yellow Aster Precambrian rocks; KD = Darrington phyllite; TG = granite intrusion (Modified from Brown et al., 1987)

Figure 16. Red Mt. (Mt. Larrabee) (right), composed of Chilliwack sedimentary rocks and American Border Peak (left) composed of Chilliwack volcanic rocks.

Figure 17. Red Mt. (Mt. Larrabee) (right), American Border Peak(center), and Canadian Border Peak (left).

Figure 18. American Border Peak (left) and Red Mt. (Mt. Larrabee) (right).

Figure 19. Red Mt. (Mt. Larrabee) in winter.

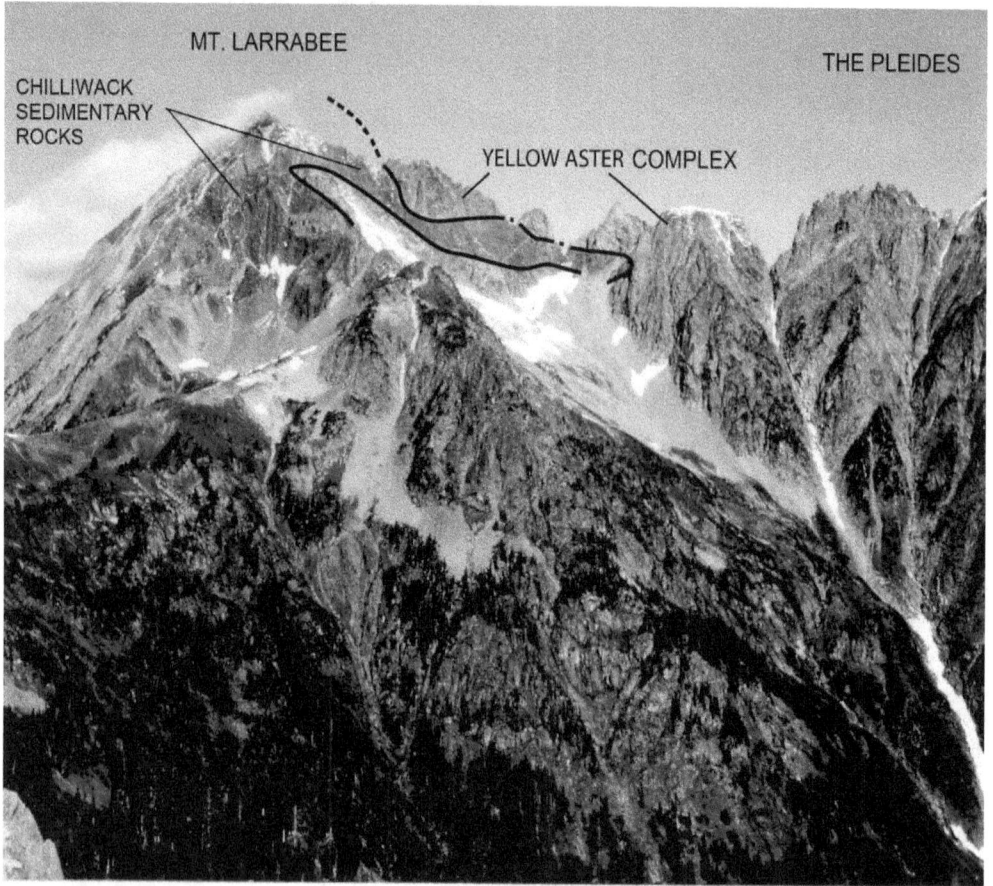

Figure 20. Red Mt. (Mt. Larrabee), a 7,868 foot peak composed of Chilliwack marine sedimentary rocks. To the east (right), The Pleides are composed of Yellow Aster granitic gneiss. Between the two peaks is a slice of Chilliwack sedimentary rocks within the Yellow Aster rocks. (Photo and geology by E. H. Brown)

Figure 21. Goat Mt. (right), composed of rocks of the Chilliwack Group.

CULTUS FORMATION—200 MILLION YEAR OLD EARLY MESOZOIC SEA

The Cultus Formation consists mostly of thin-bedded marine shale with lesser amounts of sandstone. Although it has been folded and faulted, it is not nearly as deformed as the Paleozoic Chilliwack Group rocks that underlie it. Rocks of the Cultus Formation are restricted to Sumas Mt. and do not occur elsewhere in Whatcom County.

Cultus shale contains radiolarian fossils, microscopic organisms made of silica, suggesting a Triassic age (about 200 million years ago). The extent of this Triassic sea is not known.

ELBOW LAKE FORMATION—BELL PASS MELANGE—SCRAMBLED ROCKS

The Bell Pass melange is a bit like scrambled eggs–a mixture of many bits and pieces of differing rocks, including ancient continental crust, deep ocean deposits, and pieces of the Earth's mantle. Much of the unit has been highly deformed and is made up of broken and mixed sandstone, shale, chert, basaltic lava, granite gneiss, schist, marble, and serpentine. Theses rocks have been pushed and sliced into other rocks by deep seated forces related to movement of the Earth's crustal plates.

The Elbow Lake Formation consists of sandstone, shale, chert, and basaltic lava that occur in a linear belt along the north flank of the Twin Sisters Range. Highly deformed ribbon chert contains radiolarians, microscopic fossils composed of silica, which suggest that the age of the Elbow Lake Formation is Jurassic-Triassic.

NOOKSACK GROUP—MARINE SUBMERGENCE WITH LARGE VOLCANIC ERUPTIONS 175 MILLION YEARS AGO

Wells Creek Volcanic Rocks—175 Million Year Old Volcanic Eruptions

A period of intense volcanic activity in the Jurassic Period (175 million years ago) resulted in the accumulation of submarine lava flows, volcanic breccia composed of angular rock fragments, and sandstone composed of volcanic particles. Wells Creek volcanic rocks occur at Wells Creek (Fig. 22), a tributary on the south side of the Nooksack River above the town of Glacier. Exposures of these rocks occur near Nooksack Falls and southward along Wells Creek.

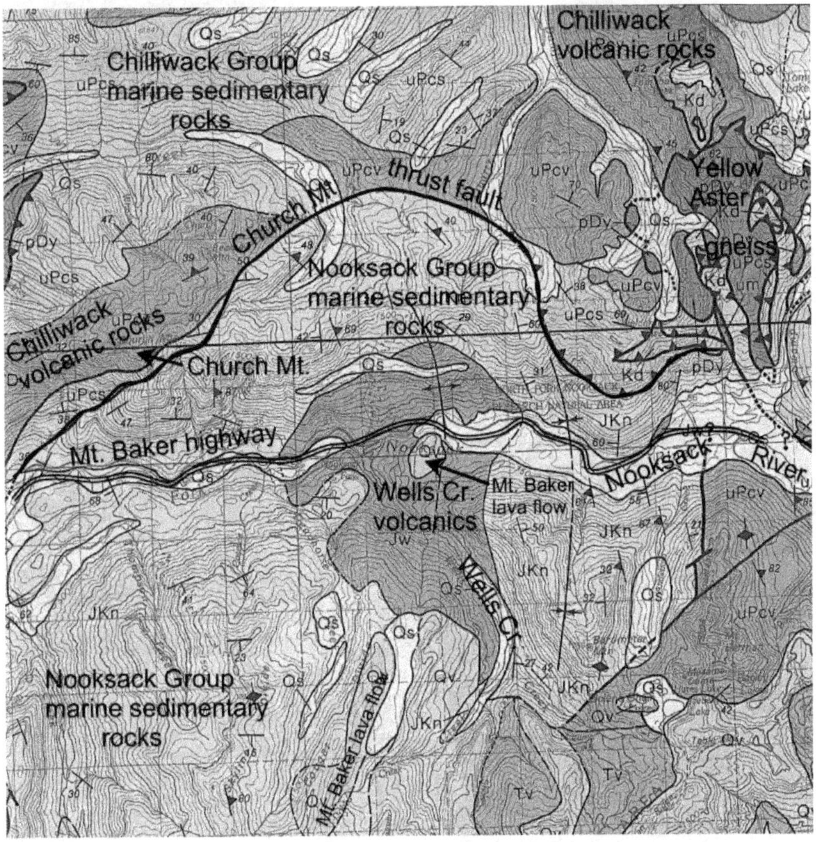

Figure 22. Geologic map of the Church Mt. area.
(Modified from Brown et al., 1987)

Wells Creek volcanic rocks occur in the core of a north-south trending anticline (Fig. 22), which brings the volcanic rocks closer to the surface where erosion has stripped off the overlying rocks.

Eight miles up valley from Glacier, the Nooksack River plunges over a cliff of resistant volcanic rocks to form Nooksack Falls (Fig. 23). The falls are a result of the difference in erodibility between the hard volcanic rocks and the relatively less resistant marine sedimentary rocks of the Nooksack Group.

Figure 23. Nooksack Falls flowing over a cliff of resistant Wells Creek volcanic rocks.

Nooksack Group Marine Sedimentary Rocks

Marine sandstone composed mostly of volcanic particles, shale, and conglomerate of the Nooksack Group overlie the Wells Creek volcanic rocks between the town of Glacier and Silver Fir campground (Fig. 22). Like the underlying Wells Creek volcanic rocks, the marine sedimentary rocks are thought to have been deposited in an oceanic volcanic arc environment perhaps similar to the modern-day Alaskan Aleutian chain.

Marine fossils are locally abundant and indicate deposition in the late Jurassic to early Cretaceous Periods (about 125 to 175 million years ago). Buchia (fossil oysters) (Fig. 24) and Belemnites (extinct nautiloids resembling squid) (Fig. 25), may be found in boulders and cobbles brought down from Church Mt. by Fossil Creek near Church Mt. Road. Thousands of feet of marine sandstone, black shale, and conglomerate that occur along Skyline Divide and Chowder Ridge in the headwaters of Dead Horse Creek just north of Mt. Baker contain fossil oysters (Buchia) (Figs. 24, 26) that are locally abundant. The fossils indicate a Jurassic age, about 175 million years ago, for the Nooksack Group rocks.

Figure 24. Fossil oysters (Buchia) in the Nooksack Group. Fossils like these near Skyline Divide above Glacier probably led to the naming of Chowder Ridge (Photo by George Mustoe)

31

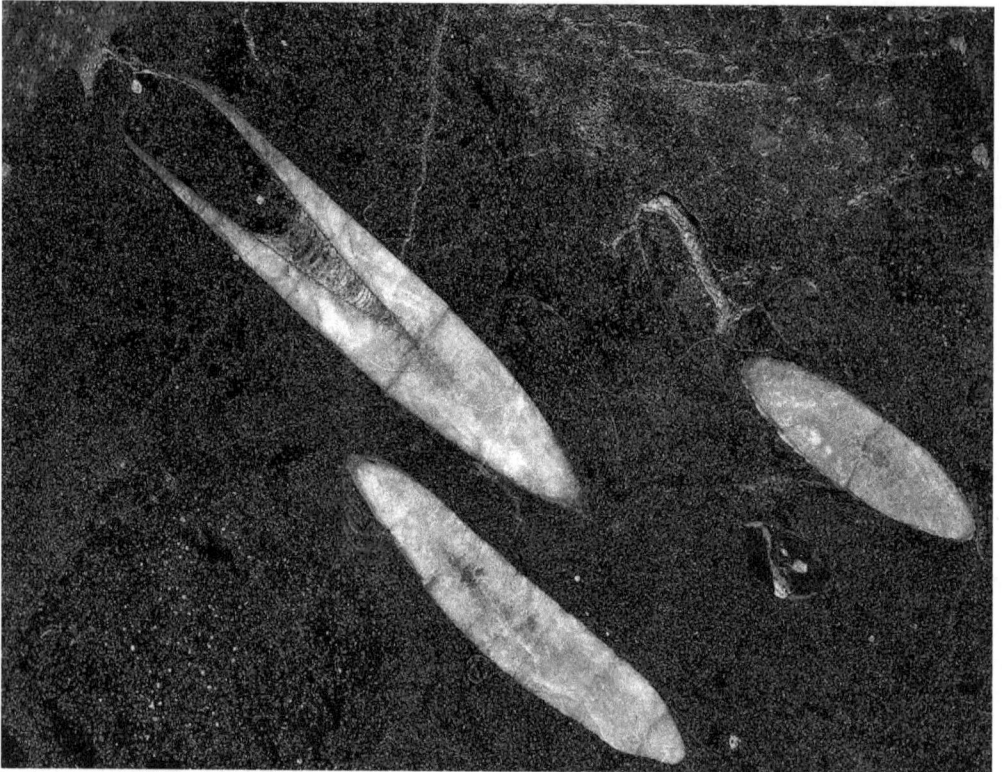

A

B

Figure 25. (A) Fossil Belemnites in Nooksack Group marine sedimentary rocks. Belemnites are an extinct marine nautiloid similar to modern squids and related to modern cuttlefish. The fossils are the shells of the belemnites. (Photo by George Mustoe)

(B) Drawing of a Belemnite as it looked when alive. They had ten arms of about equal length with small hooks for grabbing small fish and other marine animals, which they ate with their beak-like jaws. Like modern squid, they had an ink sac for defensive purposes.

Figure 26. Fossil oysters (Buchia) in Nooksack Group marine sandstone.

The Nooksack marine sedimentary rocks make up a broad area of the North Cascades between Mt. Baker and the Nooksack North Fork (Fig. 22). They extend north from Mt. Baker to the base of Church Mt. where the volcanic and marine sedimentary rocks of the Paleozoic Chilliwack Group have been thrust over them along an extensive low angle fault (Figs. 12-14), resulting in older rocks lying on top of younger rocks. After the thrust faulting, the Chilliwack and Nooksack Group rocks were folded into a broad structural arch (anticline) and the rocks at the highest part of the arch were eroded away, exposing the underlying core of Nooksack marine sedimentary rocks (Figs. 12, 22).

SHUKSAN GREENSCHIST—COOKED AND SQUEEZED LAVA 148–155 MILLION YEARS AGO

Mt. Shuksan--shuffling the crustal card deck

Shuksan greenschist was originally an oceanic basaltic lava, but it has been deeply buried and subjected to high crustal pressure and moderate heating. The resulting rock consists of fine grained, recrystallized, thinly foliated greenschist. The green color is the result of recrystallization of the original minerals in the lava to form new, green minerals—chlorite, epidote, and actinolite, which are quite small and, generally, can only be seen with a microscope. These minerals give the rocks a green color so they are known as greenschists. Isotope dating of Shuksan greenschist indicates an age of 148-155 million years (late Jurassic).

Figure 27. Mt. Shuksan reflected in Picture Lake at Heather Meadows.

Also found in the Shuksan unit are blueschists, so called because of the presence of blue, sodium–rich amphibole known as glaucophane. The mineral composition and texture of the blueschists allow us to determine the physical environment in which the rocks were formed. Like their cousins, the greenschists, blueschists form by metamorphism of basaltic lava at low temperatures and high pressures deep within the Earth's crust, approximately 15 miles below the Earth's surface where temperatures are about 700° F. At these depths and temperatures, other minerals would be expected to form, so in order for blueschists to be seen at the Earth's surface, the rocks must have been buried to great depths where crustal plates have collided and dragged basaltic lavas deep down below the surface, allowing recrystallization to occur. The blueschists were then brought swiftly back up to more shallow depths before heat could complete recrystallization of the rocks.

Figure 28. Mt. Shuksan cloaked in winter snow.

Mt. Shuksan (Figs. 27–32) consists mostly of Shuksan greenschist and blueschist thrust up to the surface along the Shuksan thrust fault, a major low angle fault (Figs. 29, 30). The Shuksan thrust fault pushed Shuksan greenschist and blueschist over the Darrington phyllite, which would normally lie on top of the greenschist. Figure 29 is a geologic map of the Shuksan thrust fault in the area between Mt. Shuksan and Mt. Baker near Heather Meadows. The Shuksan thrust fault passes between Mt. Shuksan and Shuksan Arm east of Lake Ann.

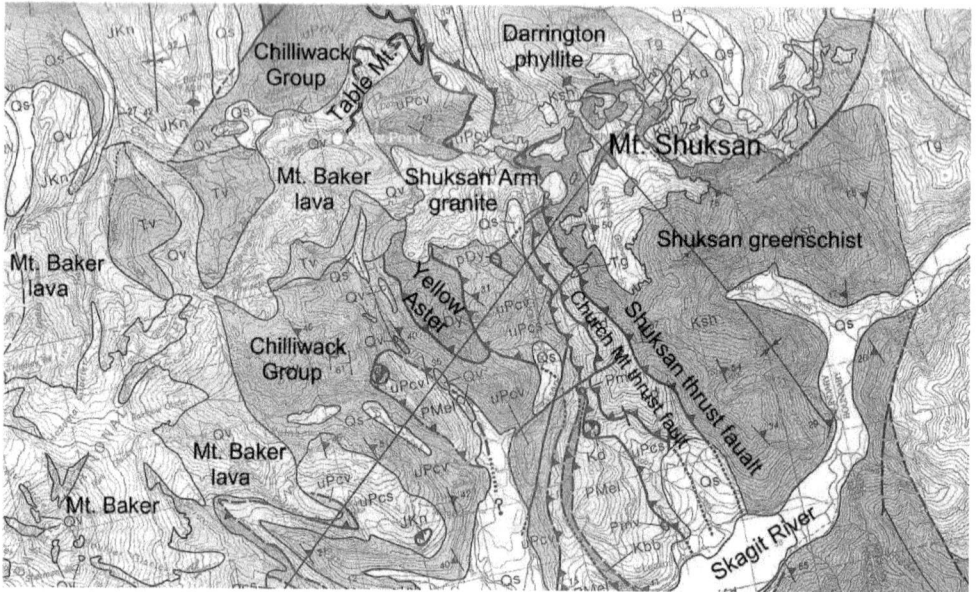

Figure 29. Geologic map of the Shuksan thrust fault zone, Table Mt., Heather Meadows, and Mt. Baker. (Modified from Brown et al., 1987)

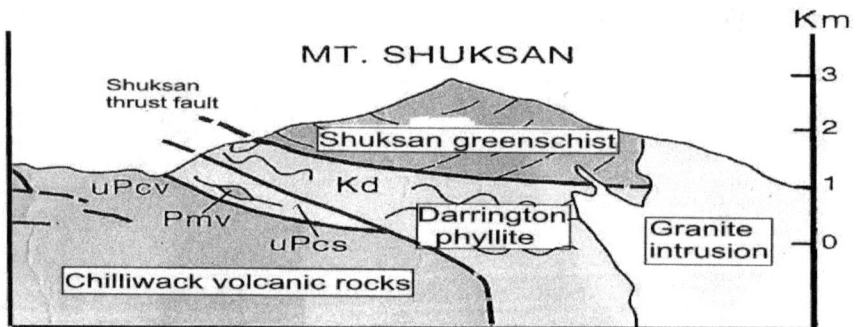

Figure 30. Geologic structure of Mt. Shuksan. The rocks here have been shuffled like a giant deck of crustal cards. Shuksan greenschist has been thrust faulted over the Darrington phyllite, which has been thrust faulted over Chilliwack volcanic rocks. (Modified from Brown et al., 1987)

Figure 31. Mt. Shuksan. the upper part of the mountain consists of greenschist that has been thrust faulted over Darrington phyllite.

Figure 32. Mt. Shuksan summit pyramid, a glacial horn carved in greenschist.

DARRINGTON PHYLLITE—COOKED AND SQUEEZED SHALE 140 TO 160 YEARS AGO

The Darrington phyllite is a gray to black, well–foliated, low grade metamorphic rock that was shale before being moderately heated and subjected to shearing pressure below the Earth's surface. The heat caused crystallization of fine–grained mica, quartz, and feldspar that give the rocks a light silvery gray sheen, even though the minerals are too small to be seen without a microscope. Shearing pressure caused the microscopic crystals of platy mica to crystallize with their flat surfaces parallel to one another , giving the rocks a foliated appearance similar to the pages of a book. Despite the crystallization, faint traces of original sedimentary bedding are commonly preserved. Prior to metamorphism, the rocks were mostly shale, siltstone, and sandstone. The thickness of the phyllite is not known, but must exceed several thousand feet and could be as much as 10,000 feet. Isotope dating indicates an age of 140-160 million years.

Figure 33. Darrington phyllite. The inclined planes in the rock are due to metamorphic foliation produced by mineral recrystallization along shear planes.

Darrington phyllite makes up a broad area south of Chuckanut Mt. and Lake Whatcom (Fig. 34). In many places it is covered with Ice Age glacial deposits but is exposed in roadcuts, creek banks, and ledges. Good examples of the phyllite may be found in roadcuts at the junction of the Deming–Sedro Woolley highway (Highway 9) and the Park road east of Lake Whatcom and at the south end of Chuckanut Drive.

Some of the phyllites are black in color as a result of much graphite (carbon) in the rock. These rocks were originally shale rich in organic material and during metamorphism, carbon in the organic material was converted to graphite, a soft, sooty-looking mineral that will make your hands black if you handle it. Because of its softness, and smearing properties, graphite is used in pencils.

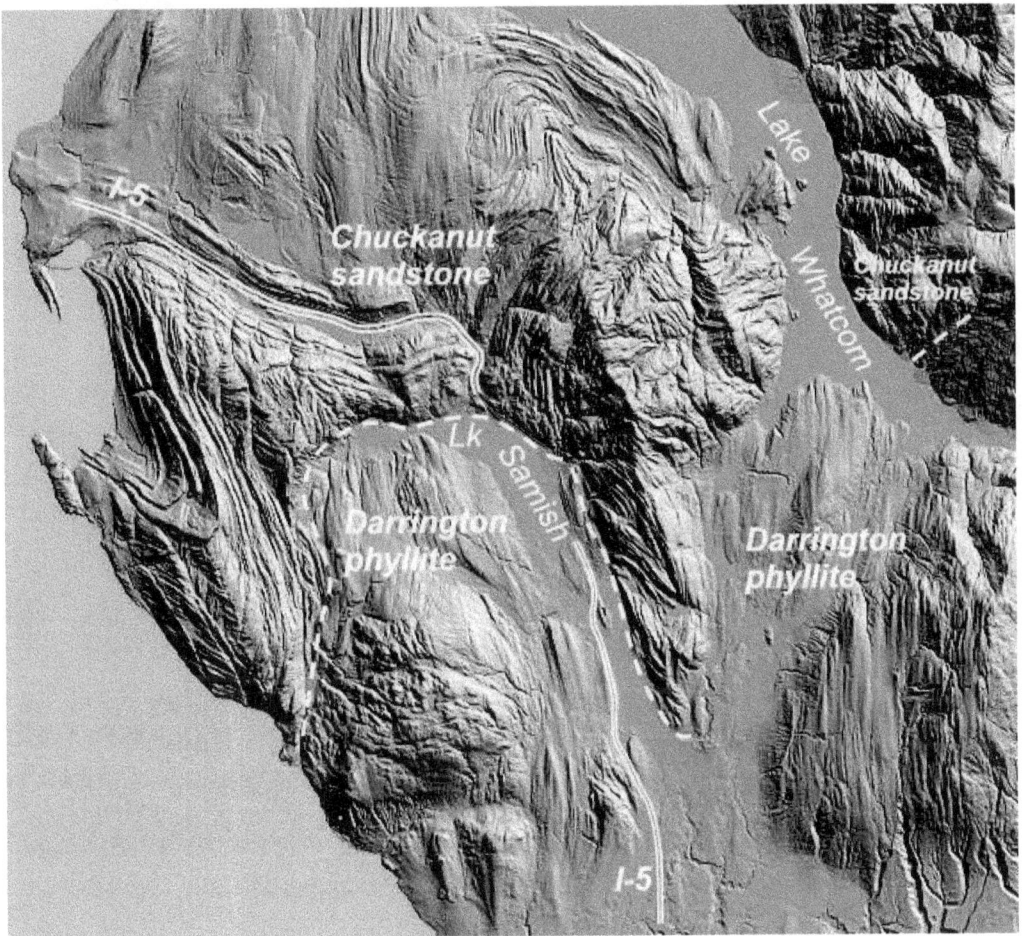

Figure 34. Distribution of Darrington phyllite. (USGS lidar image provided by Whatcom County)

Both types of phyllite are cut by many pods and lenses of quartz (Fig. 35) whose white color stands out in sharp contrast to the darker phyllite. These were formed as segregations of silica (quartz) during metamorphism. Most of the phyllite contains many blobs of white quartz about the size of a pea pod, some reaching the size of grapefruit or larger.

Minor amounts of greenschist (lava recrystallized by heat and pressure) and serpentine occur within the unit, as may be seen in roadcuts along Lake Samish and at the south end of Chuckanut Drive. The greenschist and serpentine may have been emplaced by multiple, low–angle faults or may have been depositional before metamorphism.

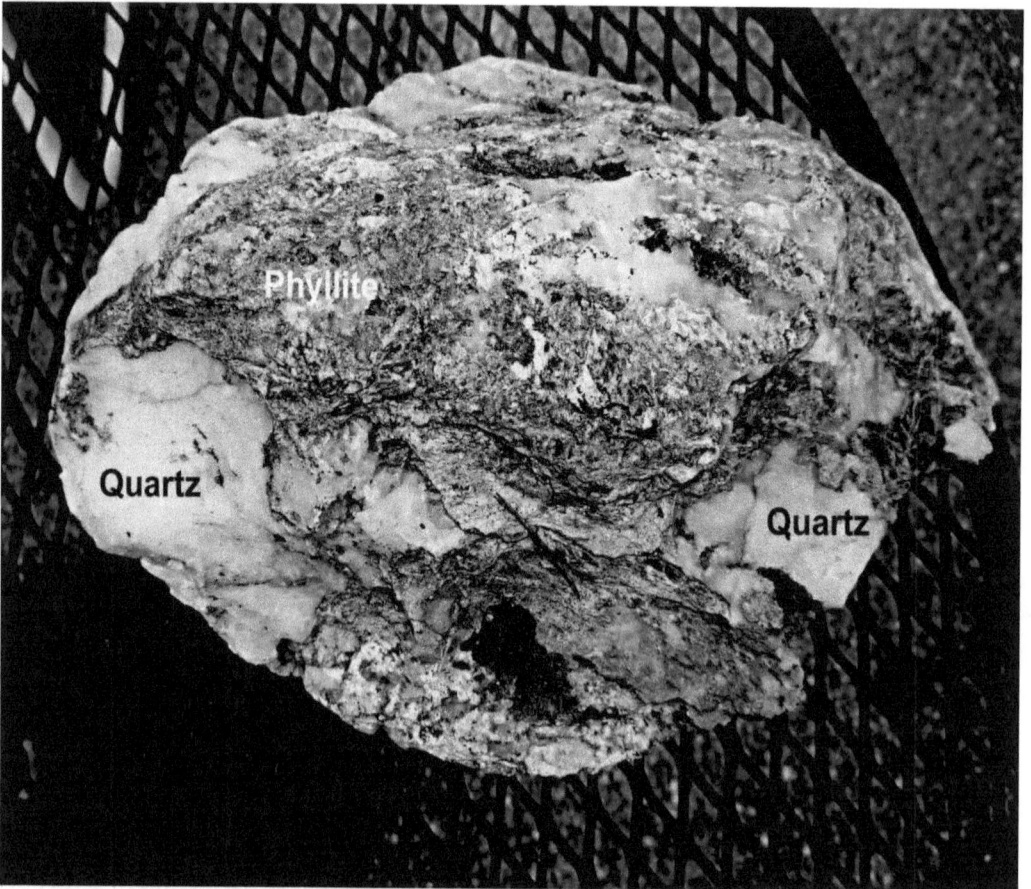

Figure 35. Quartz segregations in Darrington phyllite.

TWIN SISTERS DUNITE—
A PIECE OF THE EARTH'S MANTLE

The Twin Sisters Range (Fig. 36) is a most unusual mountain range. It is about 10 miles long and three miles wide and consists entirely of dunite, a rock composed of the mineral olivine. Olivine is a high temperature mineral found in rocks crystallized from molten magma, but entire mountain ranges of pure olivine are very rare.

Figure 36. Twin Sisters Range from Dailey Prairie.

Subsurface geophysical sounding suggests that not only is the composition of the Twin Sisters unusual, but its extent below the surface is also curious. The mass of dunite making the Twin Sisters appears to be very shallow geologically, extending only a little over a mile below the surface and not connected to anything at depth. The dunite mass lies in an area of very contorted rocks of different compositions known as a mélange. The entire mass apparently broke off from the Earth's mantle far below the Earth's crust and was thrust upward along a major fault to its present position. Thus, we have a unique opportunity to visually inspect what the Earth's mantle is made of.

41

Figure 37. Twin Sisters Range in winter.

Figure 38. Twin Sisters Range cloaked in winter snow. (Photo by J. Scurlock)

The easiest place to see the dunite, without scaling the peaks, is in boulders brought down by streams. The most spectacular examples are in the stream bed of Clearwater Creek, a tributary to the Nooksack Middle fork upstream from the Mosquito Lake bridge, where huge boulders of dunite make cascades in the creek. Smaller samples of dunite, about the size of volleyballs, may also be found at the Middle Fork bridge on the Mosquito Lake road.

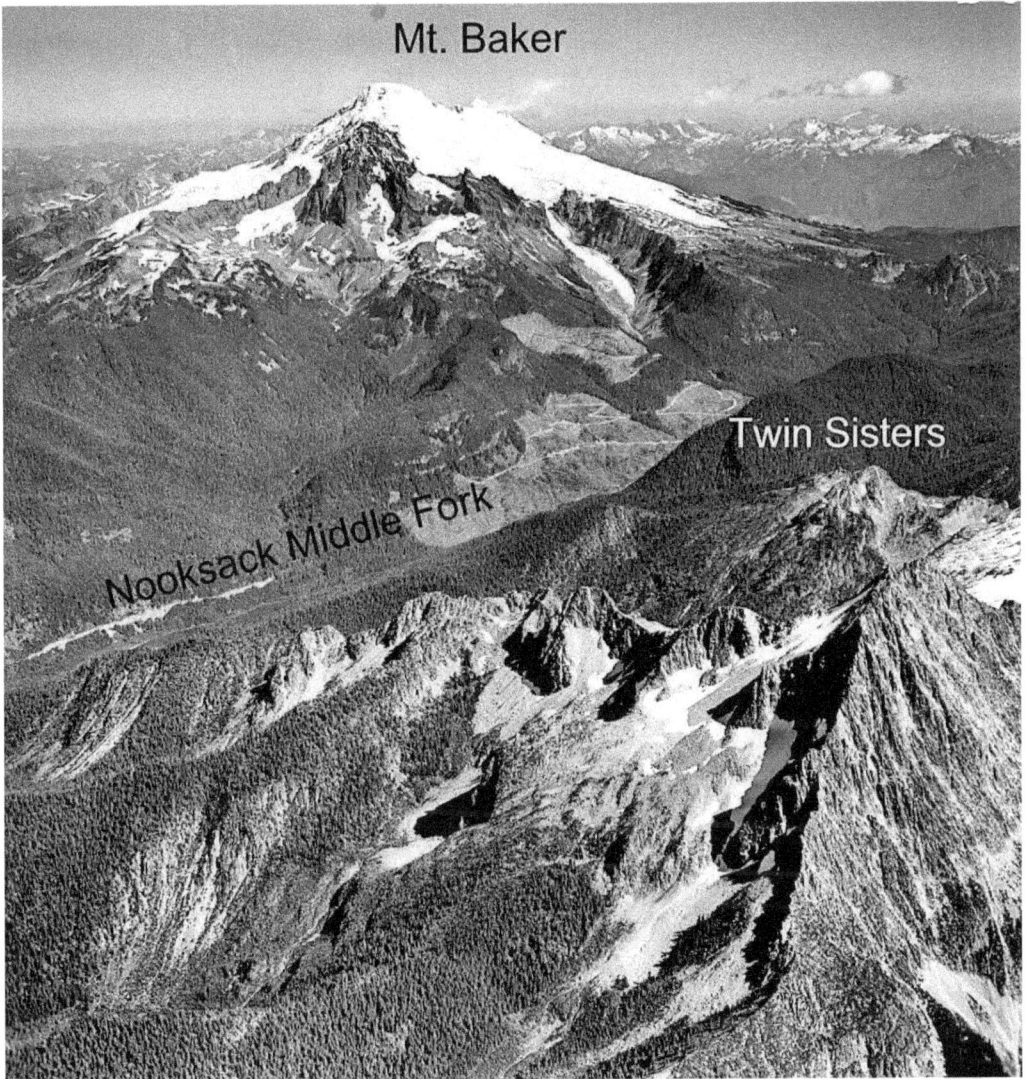

Figure 39. Looking over the summit of the Twin Sisters Range to Mt. Baker. (Photo by US Forest Service)

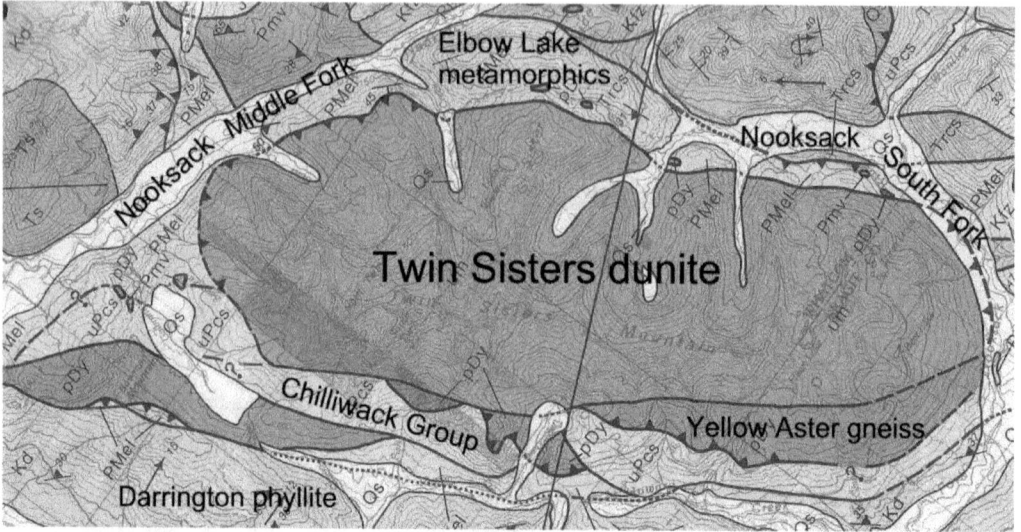

Figure 40. Geologic map of the Twin Sisters Range. The dunite making up the range is in fault contact with highly deformed rocks of the Elbow Lake Formation and the Precambrian Yellow Aster rocks. (Modified from Brown et al., 1987)

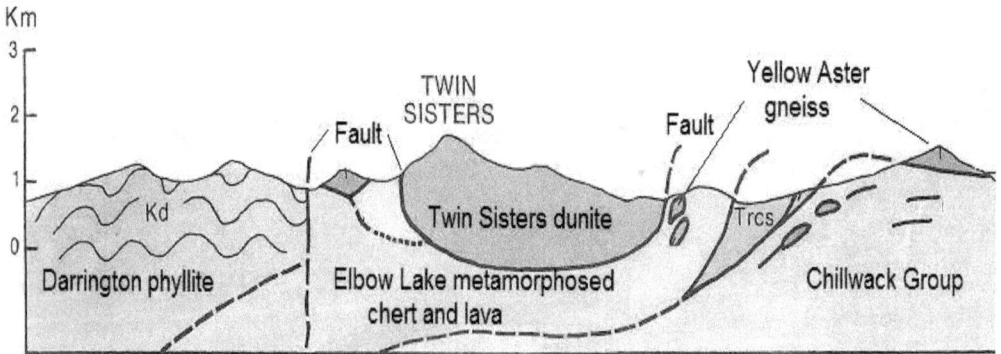

Figure 41. Geologic cross section through the Twin Sisters Range. The Twin Sisters dunite is a large rock slab that was brought to the surface along a deep fault that flattened out near the surface. The slab is bounded by highly deformed Yellow Aster gneiss and Elbow Lake metamorphosed chert and lava. (Modified from Brown et al., 1987)

Figure 42. North Twin Sister in winter.

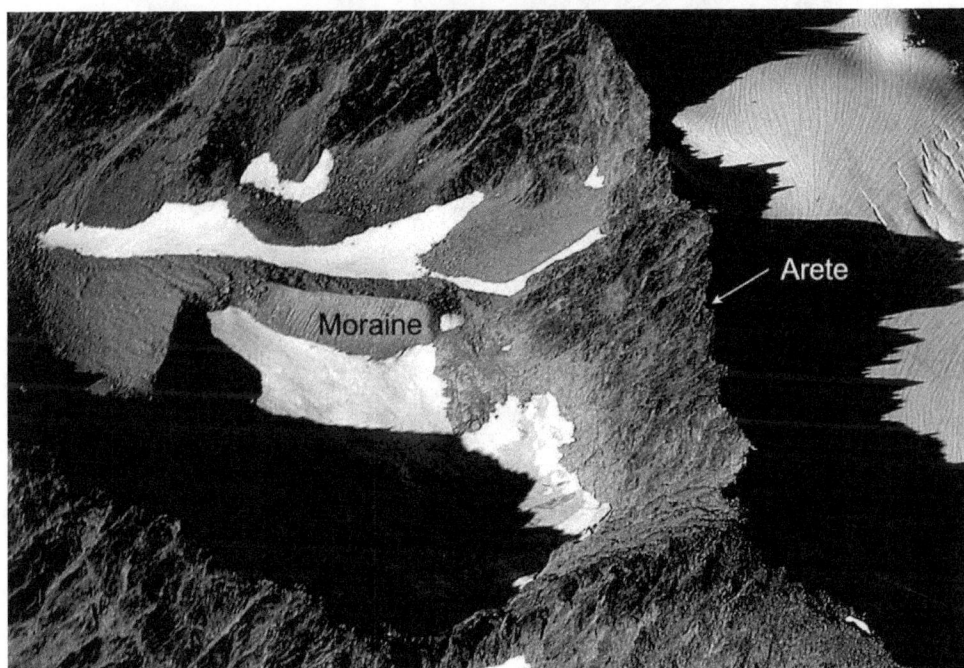

Figure 43. Moraine at the margin of a former glacier (left center) and sharp crested ridge (right) between two cirque headwalls, Twin Sisters Range.

45

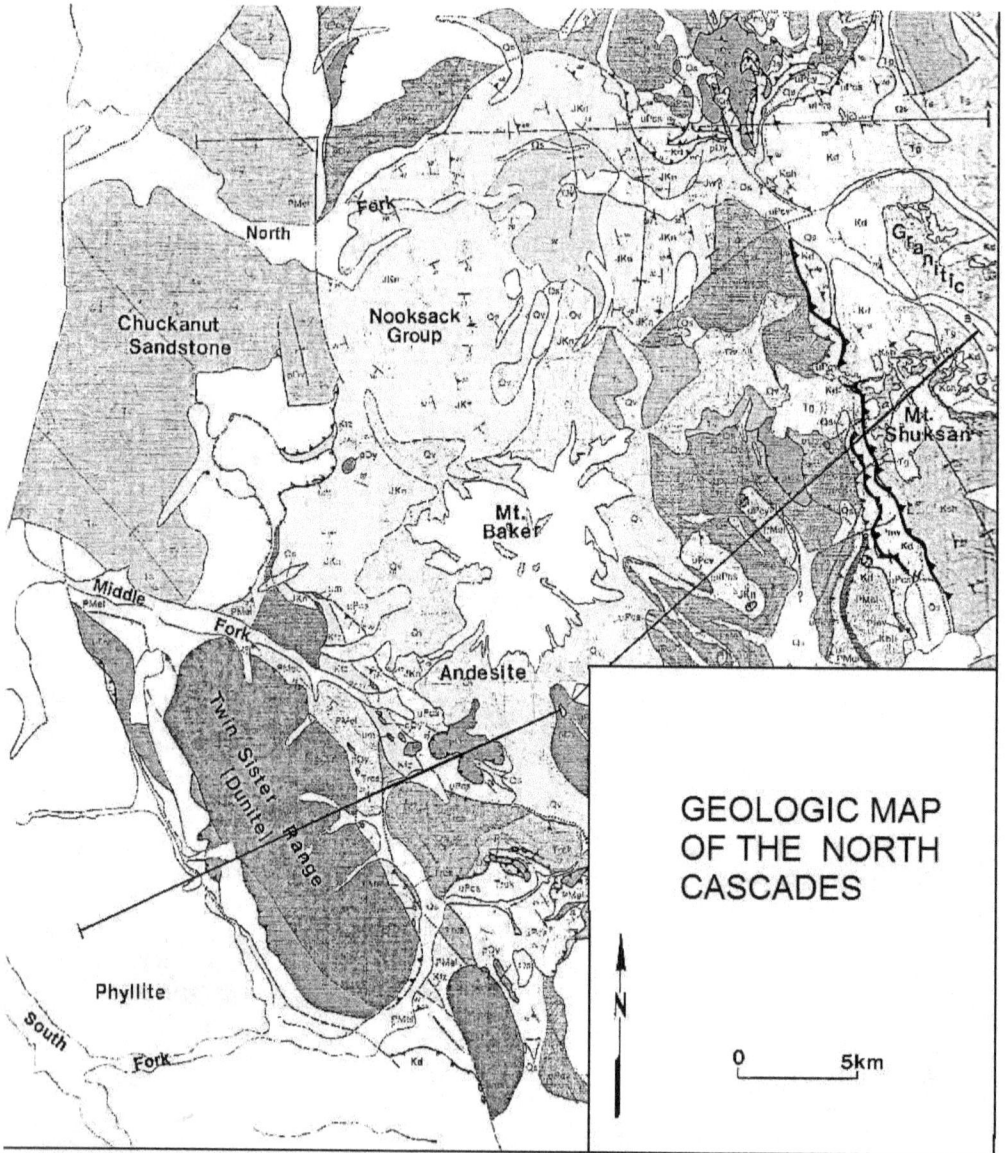

Figure 44. Geologic map of the North Cascades between the Twin Sisters Range and Mt. Shuksan. (Brown et al., 1987)

THE MYSTERY BOULDERS AT CLEARWATER CREEK

From the time I was in grade school until I left Bellingham for the university, I spent many summers fishing in Clearwater Creek pools created by huge boulders 6-13 feet in diameter (Fig. 45). When I returned many years later as a geologist, I realized that these large boulders were almost all dunite from the Twin Sisters Range on the opposite side of the Nooksack Middle Fork. The boulders in the creek bed were eroded out of a glacial deposit made up almost entirely of dunite boulders (Fig. 46), that extends 3 1/2 miles up Clearwater Creek from its junction with the Middle Fork to elevations 2,000 feet above Nooksack. The mystery is how did these dunite boulders get carried across the Nooksack Middle Fork from the Twin Sisters Range 2000 feet higher and 3 1/2 miles up Clearwater Creek on the other side of the river? The dunite boulder deposit rests on laminated lake sediments, suggesting that perhaps the overlying boulders were ice-rafted to their present location in a high, deep lake impounded by a glacier downstream. They also could have been carried there by a glacier.

Figure 45. Huge dunite boulders in Clearwater Creek.

Figure 46. Dunite boulders in a glacial deposit in the bluffs along Clearwater Creek. The large dunite boulders in Clearwater Creek were derived from this deposit.

SUMAS MT. SERPENTINE—SLIPPING AND SLIDING

SWIFT CREEK LANDSLIDE

An area of about four square miles on the northwestern part of Sumas Mt. is underlain by a mass of serpentine (Figs. 47, 48). Serpentine is a dark green mineral that forms from alteration of rocks containing the minerals olivine and pyroxene. It often has shiny, slippery surfaces that resemble polished surfaces. Shear planes and slickensides (polished surfaces made by internal slipping of parts of the rock against others) are common. Serpentine is notorious for its weakness to sliding and landslides are frequently associated with serpentine terrains.

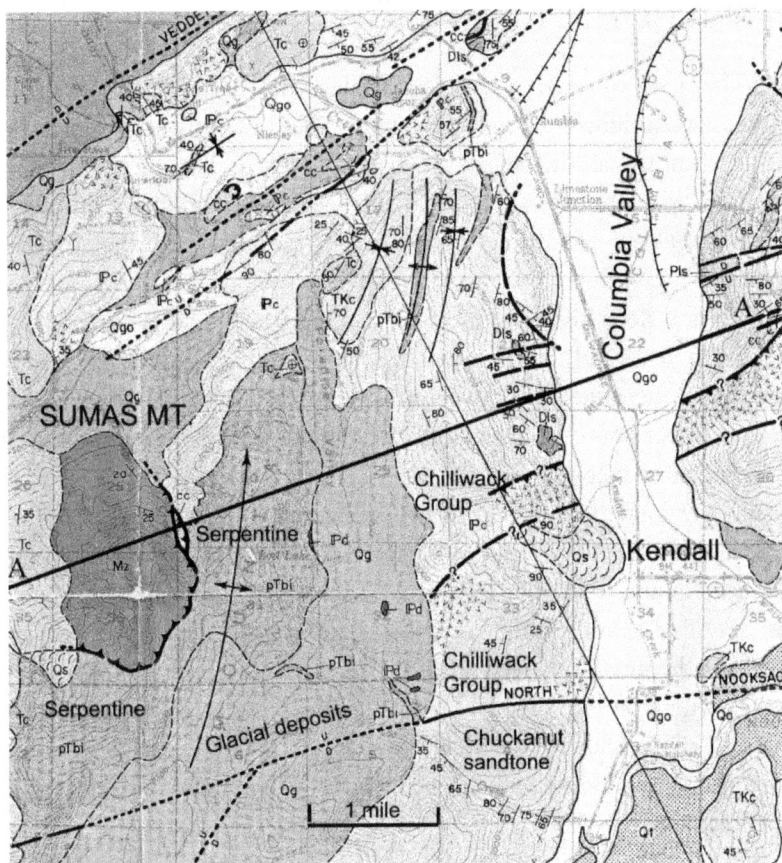

Figure 47. Geologic map showing extent of Sumas Mt. serpentine (shaded area on left) (Modified from Brown et al., 1987).

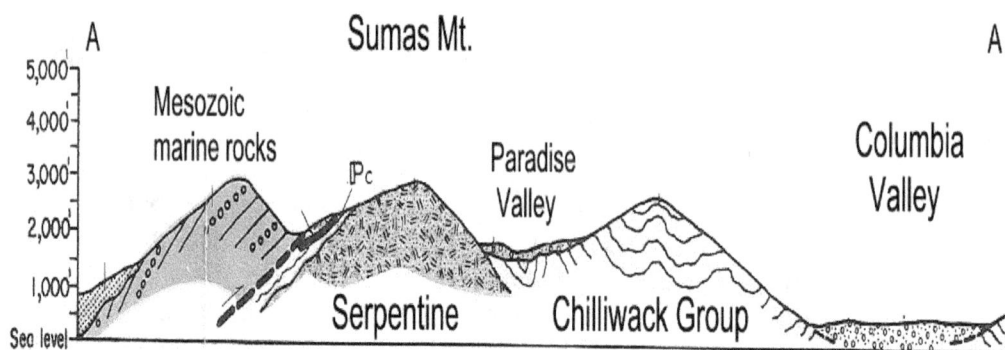

Figure 48. Geologic cross-section through the Sumas Mt. serpentine. (Modified from Brown et al., 1987)

The upper drainage of Swift Creek on Sumas Mt. starts in an area of serpentine, sandstone, and unconsolidated glacial sediments. These materials are highly unstable and a large landslide, known as the Swift Creek landslide has developed on them (Fig.49). The landslide is nearly a mile long, about 1,500 feet wide, and about 300 feet deep. It is moving at a rate of about 15–30 feet per year near the toe where the slope is steep and bare of vegetation (Fig. 49B).

The landslide is a deep-seated slope failure rooted in the serpentine, overlain by the Eocene (50 million years old) Huntingdon sandstone and conglomerate, which is broken into large blocks as the landslide moves downslope. Both are covered with glacial till. Movement in the upper part of the landslide is by block rotation, grading into more planar sliding and flowage downslope.

Runoff from the landslide carries much sediment, both as pebbles rolled along the channel floor and as a milky, muddy suspended load in the water. The very high sedimentation rate is due to erosion of the oversteepened bare slopes on the unstable toe of the landslide. For years, Swift Creek has been rapidly depositing sediment on the alluvial fan where the creek emerges from the mountain front, causing serious problems to the property owners on the fan. Numerous studies have been made of the geology of the landslide and of the feasibility of various ways of coping with the high sediment volume, but suggested solutions have been very expensive and do not offer high hopes of solving the problem.

A. The upper Swift Creek landslide. (Photo by S. R. Linneman)

B. Toe of the Swift Creek landslide. (Photo by S. R. Linneman)

Figure 49. The Swift Creek landslide on Sumas Mt.

An estimated 120,000 cubic yards of sediment is deposited on the Swift Creek alluvial fan from the landslide each year. That's equivalent to about 30,000 truckloads of sediment every year. Dredging of the stream channel to prevent the creek from flowing over roads and flooding houses has been undertaken, but has not alleviated the problem. Samples of the dredged sediment tested by the EPA in 2006 showed up to 4%, chrysotile asbestos with an average of 1.7%. Samples of dredged sediment measured by EPA in 2008 contained up to 6% chrysotile asbestos, and the Whatcom County Health Department and the Washington Department of Health jointly issued a health advisory. Heavy rain in January, 2009 caused flooding of the Sumas River and deposition of sediment containing up to 27% asbestos. Measurements of asbestos in surface water exceeded federal drinking water standards.

As sediment continues to accumulate, Swift Creek flooding becomes more common, destroying roads and burying farmland on the alluvial fan, but because of health risks and heavy expenses, little or nothing can be done to stop the flooding.

CHUCKANUT FORMATION—10,000 FEET OF 50–MILLION–YEAR–OLD SANDSTONE

Fifty million years ago, the terrain of Whatcom County looked very different than it does today. The Cascade Range had not yet been formed and the region consisted of a broad alluvial plain with streams depositing large amounts of sand in a constantly subsiding basin. Palm trees grew along the stream banks amid forests of ferns and other tropical vegetation. Streams carried mostly sand and occasionally pebbly gravel. Mud was deposited in low areas on floodplains and in shallow lakes. Extensive swamps in low areas accumulated thick organic matter that would become coal millions of years later.

The basin continued to subside as fast as the sediment was deposited for several million years until more than 10,000 feet had been laid down. Chemicals in ground water, percolating through the sand and gravel over millions of years, precipitated as cement that bound the sediment grains together to form hard rock. Mud, buried beneath thousands of feet of overlying sediment, was compressed to form shale, and the organic matter in swamps became coal. This thick sequence of sandstone, shale, conglomerate, and coal was named the Chuckanut Formation, after the rocks making up Chuckanut Mt. Most of the foothills rimming the lowland of western Whatcom County are made up of the Chuckanut Formation and may be seen along Chuckanut drive, the I-5 freeway, Sehome Hill, Squalicum Mt., Lookout Mt., the hills around Lake Whatcom, and the foothills from Deming to Glacier.

The base of the Chuckanut Formation rests on an eroded surface of Darrington phyllite. The contact between the base of the Chuckanut Formation and Darrington phyllite was exposed for some time in roadcuts adjacent to Lake Samish where a basal conglomerate of white quartz pebbles, derived from the weathering and erosion of quartz pods and lenses in the underlying phyllite, lies on the phyllite. However, these exposures are now obscured by vegetation. The basal Chuckanut sandstone at Oyster Creek along Chuckanut Drive lies on serpentine, which has been deeply weathered to form a laterite, an iron-rich soil derived from intense weathering of the underlying serpentine under a tropical climate. Both of these exposures of the basal Chuckanut show that it was deposited upon the eroded surface of the Darrington phyllite.

Figure 50. Cross-bedded Chuckanut sandstone beneath a conglomerate bed. The cross bedding is from inclined deposition of sand over the edge of a bar in an ancient stream. The numerous holes in the sandstone below the cross-bedded layer are a peculiar type of rock weathering known as honeycomb weathering. (Photo by George Mustoe)

Cross-bedding is common in the sandstone (Fig. 50). It forms when streams deposit sand on an inclined slope at the downstream end of a sand bar and results in bedding at an angle to the normal bedding. A peculiar form of rock weathering, known as honeycomb or cavernous weathering is common in Chuckanut sandstone. Decomposition of the sandstone by surface chemical weathering processes leaves the rock with multiple holes resembling a honeycomb (Figs. 50, 51). The exact process by which this type of weathering takes place is not well understood. Because it is often best seen in rock exposures along shorelines, some theories of its origin suggest it might be caused by alternate wetting from salt water spray and drying to cause precipitation of salt and other chemicals that strengthen the walls between holes. However, honeycomb weathering also occurs well inland. Fresh exposures of sandstone in roadcuts along Interstate 5 developed honeycomb weathering in only a few decades.

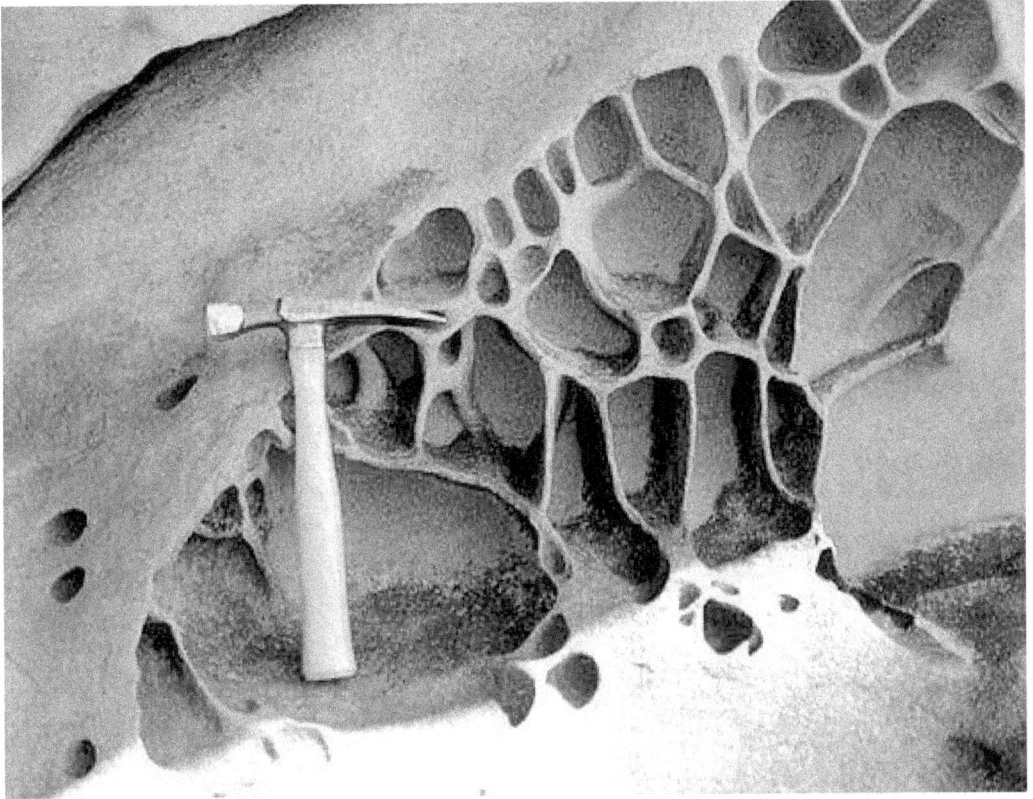

Figure 51. Honeycomb weathering in Chuckanut sandstone.
(Photo by George Mustoe)

Fossils in the Chuckanut Formation

Chuckanut sandstone is entirely nonmarine and no marine fossils have been found anywhere in it. However, plant fossils are abundant (Figs. 52-58). Leaf fossils are most common, the most interesting being palm fronds, testifying to a climate more tropical than today's. However, some scientists suggest that since continents are known to drift for considerable distances, the palm trees may have been growing much farther south than where we see palm fossils today. Fossil logs, some in growth position, occur in a few places, but are relatively rare. The fossil plants are typical of the early Tertiary period and fission-track dating of minerals in the sandstone indicates an age of about 50 million years.

Figure 52. 50–million–year–old palm frond in Chuckanut sandstone, Chuckanut Drive.

Figure 53. Palm frond in Chuckanut sandstone, Racehorse Creek. (Photo by George Mustoe)

Figure 54. Fossil palm fronds in Chuckanut sandstone. (Photo by George Mustoe)

Figure 55. Tree fern frond in Chuckanut sandstone (Photo by George Mustoe)

Figure 56. 50–million–year–old fossil tree in upright growth position. The tree was buried by sand and the wood fibers have been replaced by silica to preserve the tree.

Figure 57. Fossil leaves in Chuckanut sandstone. (Photo by George Mustoe)

Figure 58. Fossil logs in Chuckanut sandstone. (Photo by George Mustoe)

50–million-year-old fossil giant bird footprint in Chuckanut sandstone.

Figure 59. Fossil bird tracks in the Chuckanut sandstone, Racehorse Creek. (Photo by George Mustoe)

A fossil bird footprint was found in Chuckanut sandstone at Racehorse Creek in 2009 and identified by George Mustoe of Western Washington University as Diatryma, a giant, flightless, dinosaur-like bird. The three-toed footprint (Fig. 59) is 10 by 11 inches, pressed into a thin veneer of shale on top of sandstone. The bird walked across mud on top of sand, probably on the bank of a stream in a subtropical floodplain 50 million years ago. The fossil bird was probably about 6 to 7 feet tall and probably weighed about 375 to 400 pounds (Fig. 60). Other fossil bird tracks are present on the same slab of rock (Fig. 59), mostly small unknown wading birds.

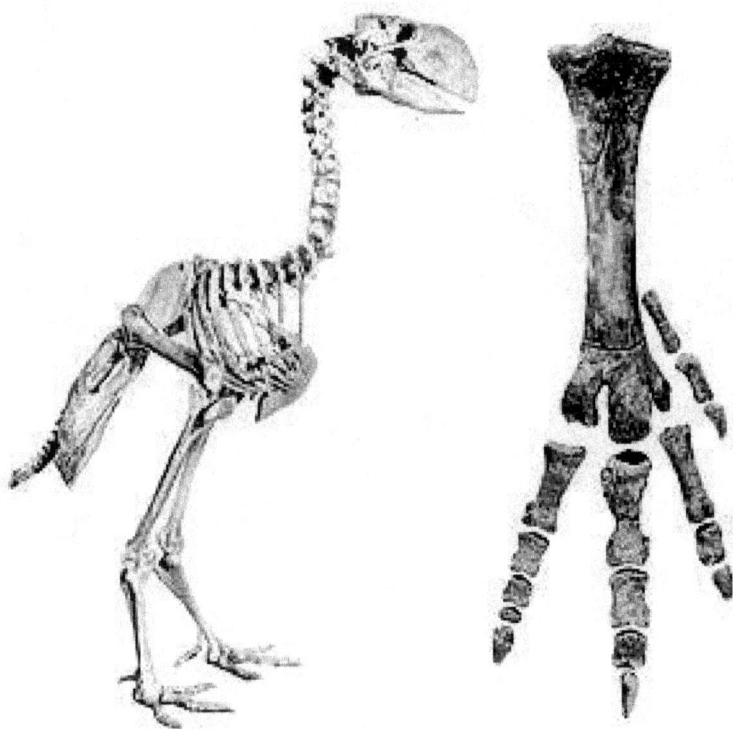

Figure 60. Reconstruction of fossil bird whose footprint was found in Chuckanut sandstone. (Museum of Natural History)

Coal in the Chuckanut

In 1852, Capt. William Pattle, an employee of the Hudson Bay Company, discovered an outcrop of coal along the eastern shore of Bellingham Bay. In 1853, he began to dig coal at the south end of what is now Boulevard Park in Bellingham. However, the quality of the coal was not good and the mine closed not long thereafter.

Figure 61. Coal bed in the Chuckanut Formation. (Photo by George Mustoe)

Sehome mine

In 1853, two loggers, Henry Brown and Samuel Hewitt, found coal in the roots of a tree near Bellingham Bay below State Street near Laurel Street. The main coal bed, interbedded with sandstone and shale, was about 15 feet thick and was tilted to the NE at about 45-70 degrees. A mine entry tunnel was started down the inclination of the coal bed for several hundred feet and expanded laterally. Coal was removed from 60'–wide rooms, leaving 20'–wide pillars to support the roof.

Brown and Hewitt took a shipload of the coal to San Francisco, found some investors, and the Bellingham Bay Coal Company was formed. The Sehome mine began production in 1855. The mine entrance began at the intersection of Railroad Avenue and Laurel Street and extended eastward beneath the area between Cornwall Street and Railroad Avenue with several offshoots, the most northerly along Prospect Street (Fig. 62). At the end of the year they had shipped 400 tons of coal to San Francisco.

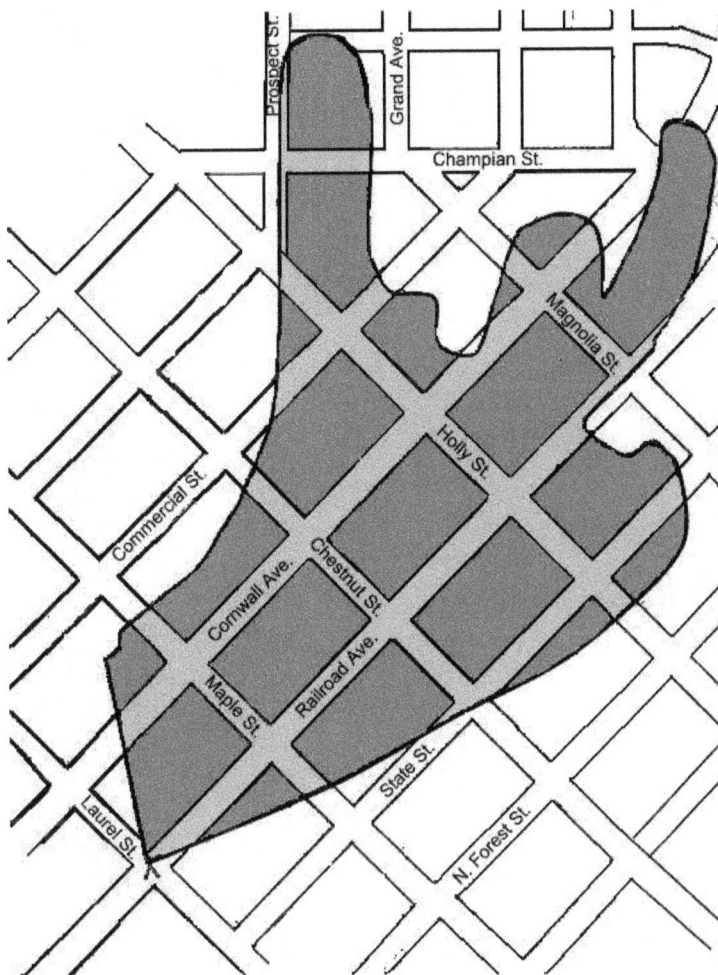

Figure 62. Underground mined out areas in the Sehome coal mine.

This coal proved to be of better quality than the Pattle mine, and the Sehome mine was quite successful, supporting about 65 miners and providing the economic base for growth of the town of Sehome along the south shore of Bellingham Bay. Ultimately, the mine produced about 250,000 tons of coal but was plagued with fires and other problems and closed in 1877.

Bellingham No. 1 mine

In 1888, Hugh Eldridge of the Bellingham Bay Improvement Company purchased 880 acres north of Squalicum Creek in North Bellingham. In 1892, ten exploration borings were made, and a coal bed about 15 feet thick was discovered 400 feet below the surface under 200 feet of unconsolidated Ice Age glacial sediments and 200 feet of sandstone.

Mining of the coal began in 1918 with a tunnel at a 30-degree slope, beginning about 400 yards northeast of the intersection of Northwest Ave. and Birchwood Ave. The slope of the tunnel was too steep, so another entrance was opened on an 18 degree slope that extended 2,000 feet to the coal bed and then another 4,500 feet down the dip of the bed.

Figure 63. Coal cars about to enter the mine portal of the Bellingham Coal Mine #1 in the Birchwood district about 400 yards east of Northwest Avenue. About 1930. (Photo courtesy of George Mustoe)

The dip of the coal bed varied from 7-18 degrees, averaging about 10 degrees. The main entry tunnel was 6,800 feet long and extended to 1,170 feet below the surface. Coal was mined from 20'–wide rooms with 30-65' pillars between to support the roof.

From 1918 to 1955, 5.3 million tons of coal was mined from the Bellingham #1 mine. The greatest production came in 1927 with 288,171 tons of coal. From 1925 to 1930, 1.5 million tons of coal was mined, averaging 250,000 tons a year. After 1930, production dropped to about half ,and from 1950 to 1955, the average annual production dropped to 59,000 tons.

When the mine closed in 1955, Bellingham was underlain by 200 linear miles of tunnels and rooms, as measured by George Mustoe of WWU from maps of the underground workings. Subsidence of the ground surface due to collapse of the mined out areas has been a problem in some areas through the years.

A long–standing myth often heard in Bellingham is that a number of Chinese miners had been killed by a cave-in and their bodies still remain in the mine. However, no evidence exists that this ever happened.

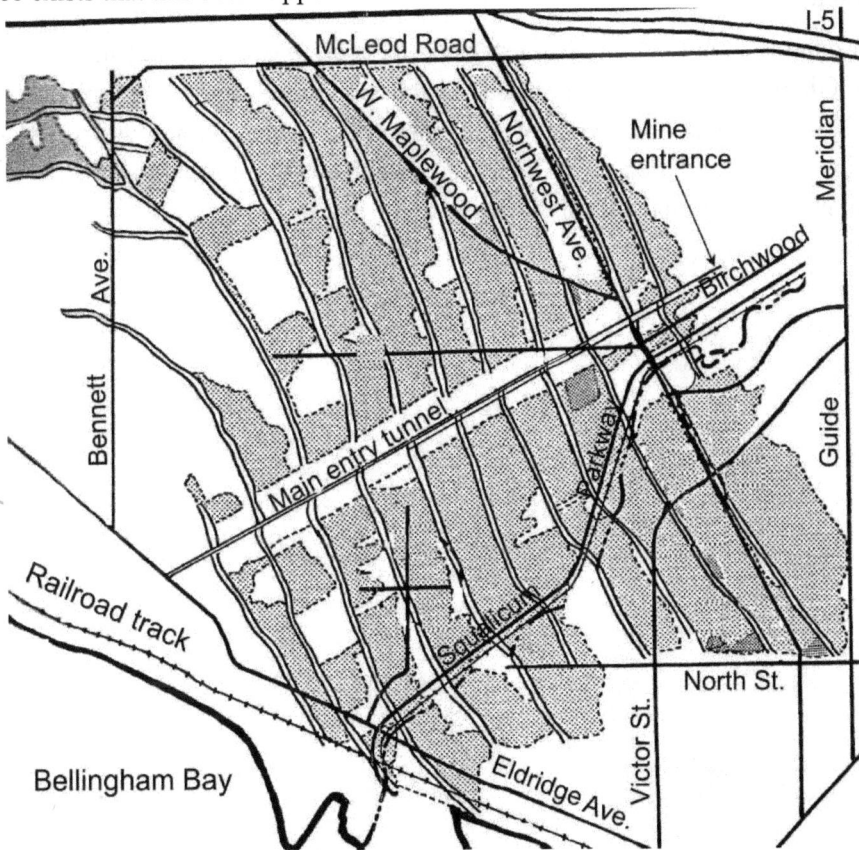

Figure 64. Underground mined out areas in the Bellingham Coal Mine #1.

The Blue Canyon mine

Coal was discovered at Blue Canyon at the southwest end of Lake Whatcom in 1887 and was mined in two separate workings: One was abandoned in 1892 because of a fault that truncated the coal bed within the basal Chuckanut Formation. However, the other mine remained in production.

The main coal bed averaged seven feet thick, but large variations were found, ranging from zero to 40 feet thick. The coal generally dipped about 30 degrees, but steepened in places to 50 degrees. The coal occurs close to the base of the Chuckanut sandstone, often lying on a thin conglomerate bed above the Darrington phyllite or directly on the phyllite.

In 1891, the mine produced 7,200 tons of coal and averaged 27,000 tons from 1892 to 1895. An explosion in the mine in 1894 killed 23 miners. After 1895, production fell to only 4,700 tons a year, except for the peak year 1900 when production reached 48,000 tons. In 1933, production was down to 333 tons, and the mine closed.

The Glacier mine

High quality anthracite, the highest grade of coal, was discovered in 1907 near the town of Glacier. All of the other coal in the Chuckanut Fm. is bituminous and the Glacier coal is the only occurrence of anthracite in the county. From 1908 to 1914, 6,000 feet of tunneling was done and by 1932, the tunnel was two miles long. However, very little coal has been extracted from the mine.

GEOLOGIC STRUCTURES IN THE CHUCKANUT— FOLDING AND FAULTING

Beds of the Chuckanut Formation have been folded into a series of anticlines (structural arches) and synclines (structural troughs) and fractured by faults. Erosion of the folds over a long period of time, measured in millions of years, has truncated the original shape of the folds, but they may be identified by the zigzag pattern of beds seen from the air (Fig. 65). Geologic structures are outlined by the erosional etching out of less resistant beds of shale and coal so that the more resistant sandstones stand out as ridges.

Chuckanut Drive crosses two such folds. Chuckanut Village lies at the axis of the Chuckanut Mt. anticline. Chuckanut Drive follows parallel to the beds on the western flank of the structure. At the sharp bend in the road near the junction with Chuckanut Point road, the beds swing sharply around, forming the Chuckanut Mt. syncline whose axis trends northwesterly toward Chuckanut Island in Chuckanut Bay (Figs. 65, 66).

Near the entrance to Larrabee State Park, sandstone beds are nearly vertical on the west flank of the Chuckanut Mt. syncline. At Chuckanut Point and Governors Point, beds dip to the northeast on the west limb of the Chuckanut Mt. syncline.

Sandstone beds making up Sehome Hill dip westward at 25-40°. Western Washington University lies in a valley of less resistant rock between two resistant beds of west-dipping sandstone.

The ridge between the Samish freeway and Lake Padden (Fig. 65) is composed of sandstone that has been tilted steeply to the north off the northern flank of the Chuckanut Mt. anticline. These beds are well exposed in road cuts along the freeway between Lake Samish and Bellingham. To the north, the beds swing around into a syncline (Fig. 65) whose axis is tilted northward. The beds on the east flank of this syncline bend around to the east, forming an anticline (Fig. 65) whose axis is tilted to the west.

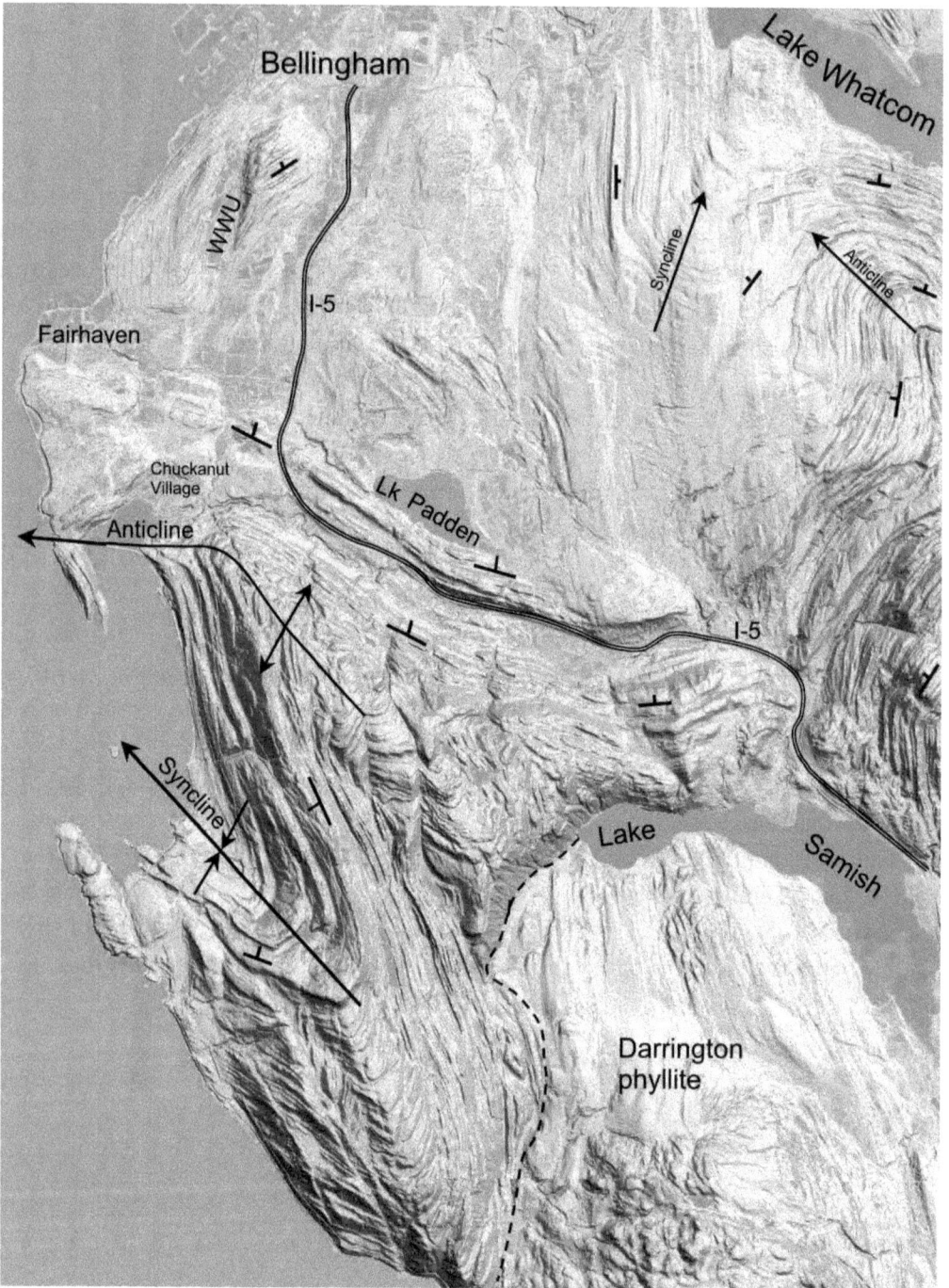

Figure 65. Lidar image of Chuckanut folds. (USGS lidar image provided by Whatcom county)

Figure 66. Chuckanut Mt. syncline and anticline. The ridges in the foreground bend into a V open to the west as part of a syncline (essentially a trough-shaped fold tilted to the west). The Chuckanut anticline lies just beyond to the north.

Structures in the Chuckanut Formation overlooking Lake Whatcom are somewhat more complicated, consisting of a number of anticlines and synclines outlined by a series of chevron-like ridges. A good example is the Strawberry Point syncline whose eastern flank lies southeast of Strawberry Point. A resistant sandstone bed on the north limb of the syncline extends into the lake, making Strawberry Point and causing the lake to shallow between Strawberry Point and its counterpart across the lake on the north shore.

HUNTINGDON FORMATION

The folding and faulting of the Chuckanut Formation that took place during the Eocene (50 million years ago) was followed by erosion that beveled many of the previously formed anticlines and synclines. A return of floodplain and deltaic sedimentation then blanketed the region with deposits lying upon structures in the Chuckanut. These rocks now make up the Huntingdon Formation.

The rocks of the Huntingdon Formation are very similar to those of the Chuckanut, mostly sandstone and shale. In outcrops, the Chuckanut and Huntingdon beds are virtually indistinguishable because their composition is so much alike. However, Huntingdon sandstone and shale which rim the foothills on the south and east margins of the lowland lie across structures which cut the underlying Chuckanut rocks. Thus structures in the Chuckanut were formed prior to deposition of the Huntingdon.

GRANITE—INTRUSIONS OF SURPRISINGLY YOUNG MOLTEN MAGMA

Granite forms deep in the Earth's crust where temperatures are high enough to melt rocks and form molten magma. The term magma is used for any molten rock. When magma is erupted on the surface, it is known as lava. The magma moves in response to crustal forces and can intrude into other rocks. As the molten mass cools, mineral crystals begin to form. The slower a magma cools, the longer the time for crystals to grow bigger, so if magma cools slowly at great depth below the surface, large crystals form. If the magma cools quickly, as in volcanic eruptions, crystals don't have time to grow large so the result is small crystals and a fine-grained rock.

Bodies of magma can be hundreds of miles long. After they solidify, they can be uplifted during mountain building episodes and the overlying rocks eroded away. Granite bodies may make up the cores of entire mountain ranges, such as the Sierra Nevada in California, the Coast Range of British Columbia, and parts of the Rocky Mts. However, the granitic intrusions of the North Cascades are much smaller and are much younger.

The Lake Ann granite makes up the eastern part of Shuksan Arm between Heather Meadows and Mt. Shuksan (Fig. 67). It was injected as molten rock into the Chilliwack volcanic rocks (Fig. 68) that make up the western part of Shuksan Arm. Alteration of the rocks that took place at the top of the granite intrusion by hot solutions of water emanating from the magma below has made the rocks bright red, as can be seen looking eastward from Artists Point toward Mt. Shuksan.

Isotope dating of the Lake Ann granite indicates that the magma had cooled and solidified by 2.75 ± 0.13 million years. This may not seem young by historic standards, but it means that the granite cooled at considerable depth below the surface, was uplifted near enough to the surface so that erosion could remove the overlying roof rocks. To do all this in 2.7 million years is indeed remarkable.

Nooksack cirque, a glacially eroded amphitheater on the north side of Mt. Shuksan, and Icy Peak on the eastern rim of the cirque (Figs. 69-72) are carved in granite that was intruded in several pulses, making several nested granite bodies. The granite in Nooksack cirque is the youngest of four phases. The oldest phase lies to the north, making the western and southern flanks of Ruth Mt. The age of these granites is not yet well known but all are probably only a little older than the 2.7–million–year–old Lake Ann granite.

The summit ridge of Mt. Sefrit is made up of similar granite (Fig. 73). Its age is unknown.

71

Figure 67. Lake Ann granite intrusion at Shuksan Arm.

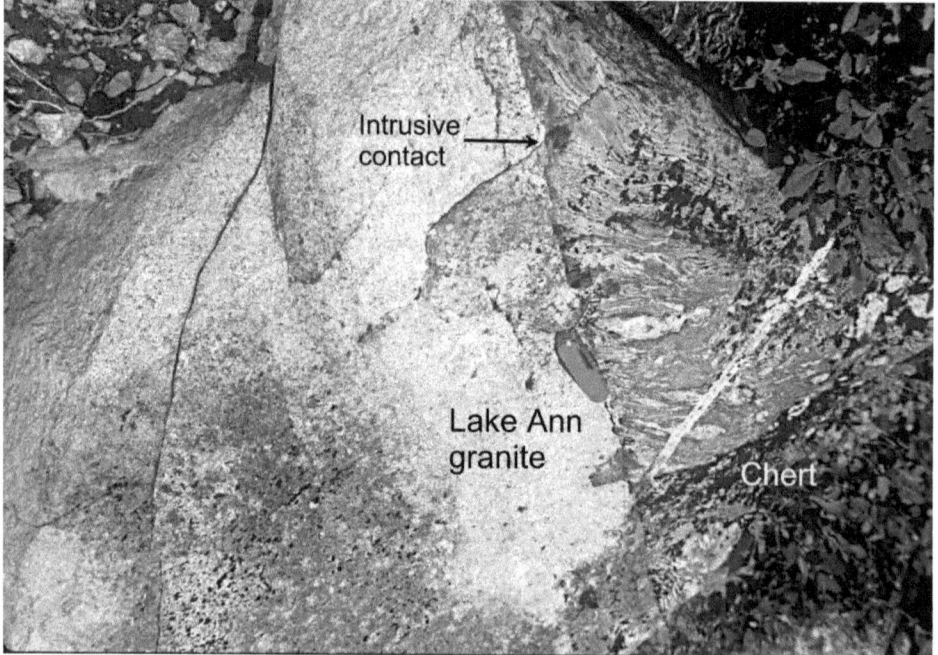

Figure 68. Contact of Lake Ann granite intruding Chilliwack chert, Lake Ann.

Figure 69. North face of Mt. Shuksan composed of young granite and Darrington phyllite. (Photo by J. Scurlock)

73

Figure 70. Cirque carved in Darrington phyllite intruded by granite on the north face of Mt. Shuksan above Price Lake. (Photo by J. Scurlock)

Figure 71. Icy Peak (left) and the headwall of Nooksack cirque carved in granite.

Figure 72. Headwall of Nooksack cirque made of young granite.

**Figure 73. Mt. Sefrit from Heather Meadows.
The summit ridge is composed of granite.**

CALDERAS—CRATER LAKE COUSINS

Calderas are the roots of ancient volcanoes. Most, like Crater Lake in Oregon, are characterized by very explosive volcanic eruptions, followed by collapse of the volcano to form a caldera. They differ from normal volcanoes in lacking a high central peak and usually consist of a broad, circular basin filled with volcanic rocks.

In the headwaters of Swift Creek east of Table Mt. just south of Artists Point, deposits of white volcanic ash and pumice are exposed in gullies leading down to Swift Creek. These deposits lie in a roughly circular basin known as the Kulshan caldera. The thickness of the ashy deposits is estimated at several thousand feet on the basis of exposures in gullies descending into Swift Creek. However, the sides of the caldera volcanic vent are not exposed.

Figure 74. Pumice and ash flows of the Kulshan caldera in the center of the photo. Looking south from Artists Point toward Ptarmigan Ridge. Table Mt. and the Chain Lakes trail are just off the photo to the right.

Many years ago (1981), I discovered a thick volcanic ash 8–12 inches thick in the Puyallup Valley south of Seattle. The ash, now known as the Lake Tapps ash, consisted of fine-grained pumice and volcanic glass. We determined the chemical composition of the ash by x-ray fluorescence, allowing us to correlate it from Sumner and Auburn in the Puyallup Valley to Hoods Canal on the Olympic Peninsula. An age of one million years was measured by fission track analysis. The chemical composition didn't match any of the volcanoes of the Cascade Range, so we didn't know where it came from. Recently, the ash deposits of the Kulshan caldera were found to be very similar in chemical composition to the Lake Tapps ash in the southern Puget Lowland. An isotope date of 1.1 million years strongly suggests that the source of the Lake Tapps ash was the Kulshan caldera. The eruption that sent so much ash so far south from the Kulshan caldera must have been very explosive, probably similar to the eruption of Mt. Mazama at Crater Lake.

Figure 75. Lake Tapps ash at Auburn, south of Seattle, derived from an explosive eruption of Kulshan caldera.

MINING--THAR'S GOLD IN THEM THAR HILLS!

The discovery of gold on the Fraser River in British Columbia set off a gold rush as prospectors and miners flocked to the area. The settlements around Bellingham Bay saw opportunities for becoming a center for outfitting prospectors on their way to the Fraser River and undertook plans to build a trail north to connect with the Hudson's Bay Company's Brigade Trail up the Fraser River. While awaiting completion of this trail, prospectors began looking for gold in the Nooksack drainage. The first discovery of gold in Whatcom County was made July 24, 1858, when William Young arrived in Whatcom with two gold nuggets found 11 miles northeast of the mouth of the Nooksack River.

Most local prospectors joined the Fraser River stampede when the Whatcom trail to the Fraser and Thompson River areas of British Columbia was finished. However, a party of Whatcom prospectors found placer gold on the South Fork of the Nooksack River in 1860. In the late 1870s, a man named Rowley discovered placer gold in Ruby Creek in the eastern part of the county. This set off a rush of several hundred prospectors and more than $100,000 in placer gold was reportedly mined from stream gravel. Placer mining continued with nuggets found up to an ounce in weight, and this set off searches for the 'mother lode.' In 1891, the first lode claim, the Nellie Belle, was staked by Franklin Rives, followed soon thereafter by a claim staked by Henry Benke and partners near the headwaters of Canyon Creek. In 1893, the Eureka lode was discovered by A.M. Barron, setting off a gold rush in the Slate Creek mining area that resulted in 3,000 claims.

Small scale placer mining continued on the South Fork of the Nooksack River and by 1885, several hundred prospectors had staked out the river banks and established the town of Livewood near the mouth of Skookum Creek, a tributary to the Nooksack South Fork. However, the amount of gold there was not sufficient to sustain operations.

In the summer of 1894, E.H. Thomas and J.W. Hulett staked several gold-silver claims and by 1895, a wagon road had been built from Whatcom to Cornell Creek near Glacier. A trail, known as the Cascade State Trail, extended another 20 miles to the east to within two miles of Hannegan Pass.

The first significant discovery of lead, zinc, and silver was made in 1896 by H.C. Wells on Ruth Creek near the end of the Cascade State Trail in the upper North Fork of the Nooksack. Several other occurrences of gold-bearing quartz were found in the surrounding area.

The most important discovery of gold was made on August 23, 1897 when Jack Post of Sumas found the Lone Jack vein on the south side of Bear Mt. above Silesia Creek while prospecting with Russ Lambert, and L.G. Van Valkenburg. Assays of the gold ran as high as $10,750 a ton and set off a gold rush in the area. Claims were staked in the area around Twin Lakes near the Lone Jack mine and on Red Mt. (later renamed Mt. Larrabee), Tomyhoi Peak, Goat Mt., and upper Silesia Creek. In the summer of 1898, C.W. Roth and others filed claims on the north side of Red Mt. about half a mile south of the Canadian border. The Red Mt. Gold Mining Company was formed in 1900 and additional claims were staked. During this time, placer claims were also being staked along the Nooksack River and in Ruth, Swamp, and Silesia Creeks, but yielded only small amounts of gold. Also in 1900, G.B. Conway discovered gold, silver, and copper on Damfino Creek four miles west of Twin Lakes, and the Mt. Baker-Shuksan Mining Co. drove tunnels there in 1904.

In 1901, William Boyd and W.L. Martin filed a claim on the south side of Red Mt., which later became the Gargett mine, and L.A. Price and others staked claims northeast of Twin Lakes. Gold and silver were discovered on Wells Creek, which flows into the Nooksack River at Nooksack falls, in 1901. By 1903, the Great Excelsior Mining Co. had set up a stamp mill and shipped gold and silver to the Tacoma smelter until 1916. Another stamp mill produced gold-silver concentrates at the Nooksack mine near Sumas but closed down in 1908. In 1904, the Blonden brothers filed a claim near the headwaters of Swamp Creek.

South of Mt. Baker, claims were staked along Swift, Shuksan, Sulphide and other creeks flowing into the Baker River. In 1908, Joe Morovits filed a claim in upper Swift Creek and moved a stamp mill there, but the amount of ore was not enough to keep the mill running and mining ended in 1912. From 1890 to 1937, more than 5,000 mining claims had been filed.

Lone Jack Mine

In 1896, Jack Post, a veteran prospector from Sumas, found gold in Swamp Creek, a tributary to the Nooksack North Fork near Silver Fir Campground, and followed it upstream to its source near Twin Lakes. He also found promising quartz boulders in Silesia Creek. Post was joined by Russ Lambert (my great uncle) and Luman Van Valkenburg, a logger, in their search for gold. In 1897, Post proposed that they investigate the area around Twin Lakes and had been camped there for over two weeks when Post suggested that they explore separately the next day, August 23, 1897, and meet back at camp at the end of the day. Lambert and Van Valkenburg were the first to return to camp and as evening approached, they had begun to worry about what might have happened to Post when they heard him yelling "I've found it, I've found it!" He had indeed discovered the richest gold strike in the Pacific Northwest, a gold-bearing quartz vein on the southeast side of 6,440-foot Bear Mtn. (Fig. 76)

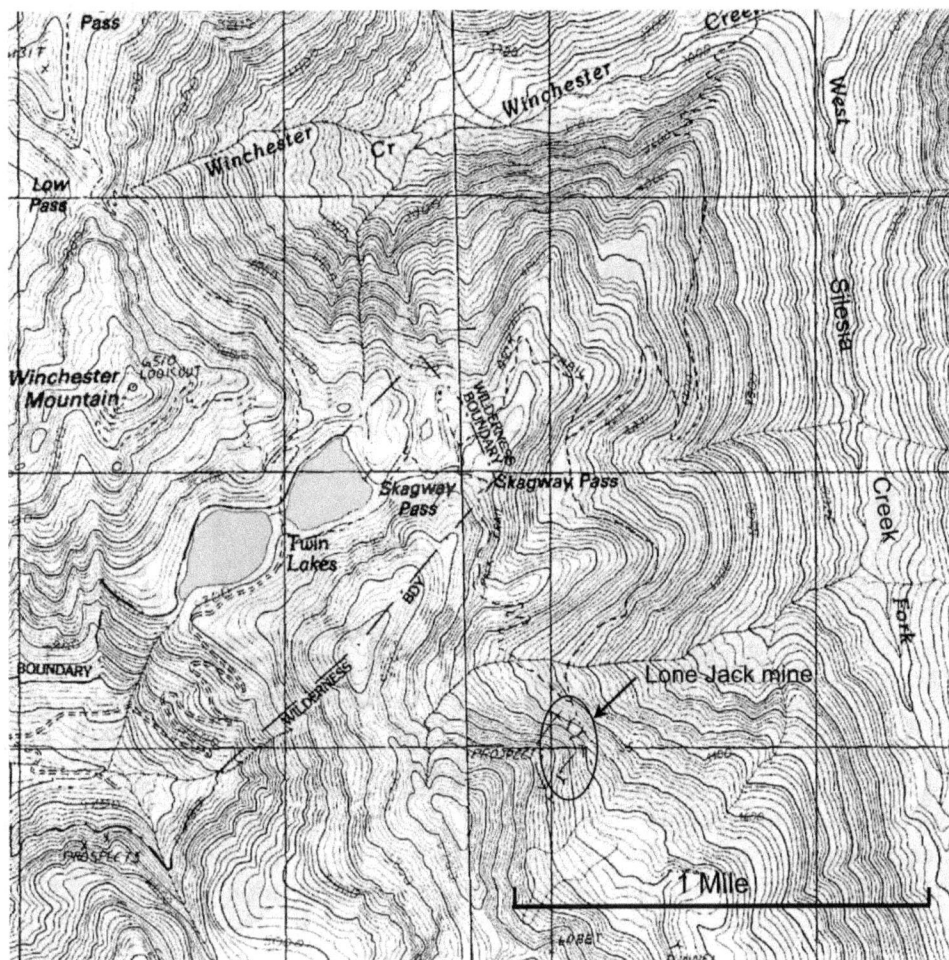

Figure 76. Map of the Lone Jack mine. (USGS topographic map)

The three had agreed to share equally whatever any of them found and they staked five claims. Post got the discovery claim, the Lone Jack, and Lambert and Van Valkenburg each got a claim. Two joint claims, the Whist and Lulu, were filed in all three names. The Lulu lode later turned out to be one of the richest. Lambert went back to Sumas to file the claims and submit samples for assay. When the assays came back, they were an astounding $10,750 per ton. When word of this spread through Sumas, it immediately set off a gold rush in the area.

81

Gold in the Lone Jack mine occurs in quartz veins that formed by precipitation of minerals from hot water percolating through fractures in the rocks. The source of the hot water carrying dissolved minerals was from nearby granite intrusions. Contact of the hot water with the cool sides of fractures causes minerals to precipitate along the sides of the fractures to form veins. If gold is present in the hot solutions, then gold-bearing quartz forms.

Figure 77. Gold ore from the Lone Jack mine. The white mineral is quartz and the darker mineral is pyrrhotite, an iron sulfide mineral frequently found in ore deposits. The gold is too fine to see without magnification. (Photo by George Mustoe)

The host rock for the quartz veins is black Darrington phyllite, which contains many small veins and lenses of older quartz not related to the gold-bearing veins. The gold-bearing quartz consists of two distinct generations; (1) older, white, coarse grained quartz that has been micro-fractured, and (2) fine-grained, younger quartz containing free gold that has cemented the earlier ground up quartz. Gold in the quartz is mostly too small to see without magnification, although some parts of the vein contain pinhead-size specks of gold. Gold has been mined from three principal quartz veins, the Lone Jack, Lulu, and Whist, which may be faulted segments of the same vein or separate veins emplaced under similar conditions.

Figure 78. View to the south at Lone Jack mine site. (1) trail to Lone Jack portal, (2) flotation mill site, (3) Lulu portal, (4) Whist vein outcrop, (5) Lulu vein outcrop. (From Wolff et al., 2005)

83

The Lone Jack vein (Fig. 79) is a massive quartz vein about three feet thick in Darrington Phyllite between two high-angle shear (fault) zones. The vein strikes N10°W and dips 45 degrees to the west. It pinches out at the south end and is truncated 500 feet north by a fault.

The Lulu vein (Fig. 80) is parallel to foliation in Darrington phyllite and varies in width from several inches to 9 feet. It strikes generally east and is inclined 8° to 60° to the south.

The Whist vein lies within a fault that truncates the Lone Jack vein and pinches out upward into this fault. The vein strikes S10°W and is nearly vertical. It is exposed for 80 feet and then disappears under accumulations of loose rock.

Figure 79. Entrance to the mine of the Lone Jack vein. (From Wolff et al., 2005)

Figure 80. Entrance to the mine of the Lulu vein. (From Wolff et al., 2005)

Mining operations at the Lone Jack claims occurred in several discontinuous periods, 1900–1907, 1916–1917, 1922–1924, 1991–1996, and at present. After discovery of the Lone Jack in 1897, the claims were sold to Henry Hahn and Leo Friede for $50,000 in 1898, and the Mt. Baker Mining Co. was organized. In 1900, a 10-stamp mill was hauled by steam donkey and horses over a trail from Glacier to Twin Lakes and lowered down a switchback trail 4,000 feet from the mine to a place near Silesia Creek. In 1901, a 50-ton aerial tram was installed between the mine and the mill, and five more stamps were added to the mill. The Mt. Baker Mining Co. mined the Lone Jack lode from 1898 through 1907, during which time most of the gold and silver were mined. In 1907, the stamp mill burned, the aerial tramway collapsed, and operations ceased.

No mining took place from 1907 to 1915. In 1915, the property was leased to the Boundary Gold Co. and a 10-foot grinding mill was built on the hillside above the old mill. After only several hundred tons of ore had been mined from the Lulu vein, operations ended in 1917.

The Brooks-Willis Metals Co. leased the mine in 1917 and purchased it in 1919. In 1923, a 100-ton flotation mill, bunkhouse, sawmill, and flume, were built on the hillside below the Lulu portal and a hydroelectric plant was built on Silesia Creek for power. Ore from the Lulu lode was mined in 1923, but operations ceased when avalanches destroyed the flotation mill in the winter of 1924/1925.

U.S. Forest Service officials became alarmed that significant quantities of nitroglycerin had leaked out of several hundred cases of dynamite that had been stored in the Lulu drift since the 1920's, and, in August 1964, detonated it, destroying the mine tracks and air lines but not damaging the mine itself.

In 1992, the Diversified Development Co. began mining the Whist vein, which crops out about 800 feet northwest of the Lulu portal and several hundred feet above it. In 1995, the Whist vein was reached from a new tunnel below the outcrop. From 1992 to 1996, about 800 tons of ore per season were shipped to the Asarco smelter in East Helena, Montana. Activity at the mine is currently continuing.

U.S. mint receipts show 10,370 ounces of gold and 2,000 ounces of silver were mined from the Lone Jack veins. The total dollar value adjusted to today's precious metals market would amount to $13.5 million.

**Figure 81. Flotation mill and bunkhouse below the Lulu lode.
(From Wolff et al., 2005; Whatcom County Museum)**

Boundary Red Mine

The discovery of gold at the Lone Jack lode resulted in a rush of prospectors into the Twin Lakes area. Among them was Thomas Braithwaite. Having found no gold and in need of food, he and his partner went goat hunting on Red Mt. (Mt. Larrabee). They shot a goat, which fell into a deep ravine, and when they found the goat, it was lying near a ledge of gold-bearing quartz that later became the Boundary Red Mine, the second richest mine in the region. However, the mine was in a very remote area about half a mile south of the Canadian border and three miles north of the Lone Jack mine (Figs. 82, 83).

In 1900, the Red Mt. Gold Mining Co. was formed and mining began. By 1913, a stamp mill was crushing ore and in 1914 the Boundary Red Mt. mine was the leading producer of gold in the county. The mill was built on a narrow patch of forest between

86

two avalanche chutes. Ore was delivered to the mill via an aerial tramway. Power to the mill was provided by a generator on Silesia Creek via a 1.5-mile-long transmission line. Because of the remoteness of the mine, steep terrain, and difficult conditions, maintaining a work force was not easy. As one observer noted, "Labor turnover was so rapid that it required the proverbial three crews—one coming, one working, and one leaving—to keep the mine in operation. Not uncommonly, men arrived and departed without having worked one full shift."

Five additional stamps (ore crushers) were added to the mill in 1916 and 10,440 tons of ore were processed, bringing in $148,578. In 1917, $132,000 in gold was mined, but operations were suspended because of fire and World War I. In 1918 the power plant on Slesse Creek was destroyed by fire. Mining resumed in 1921 and from 1923 to 1930 an average of about $75,000 per year in gold was mined. Beginning in 1931, gold production diminished sharply, averaging about $8,600 from 1931 to 1942 when the mill was destroyed by a snow slide. From 1913 to 1946, the total gold and silver production was $947,579.

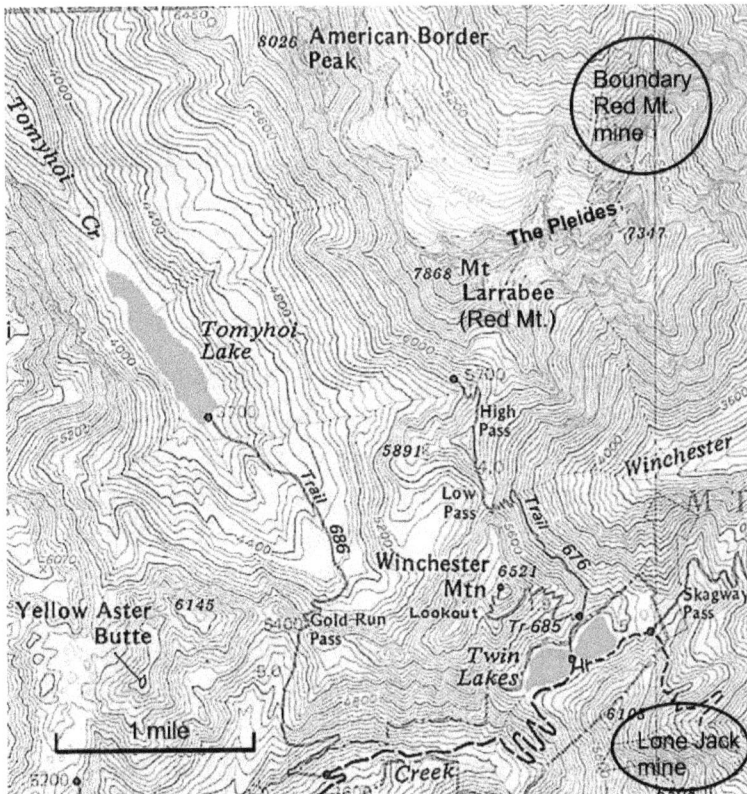

Figure 82. Topographic map of the Boundary Red Mt. mine area. (USGS topographic map)

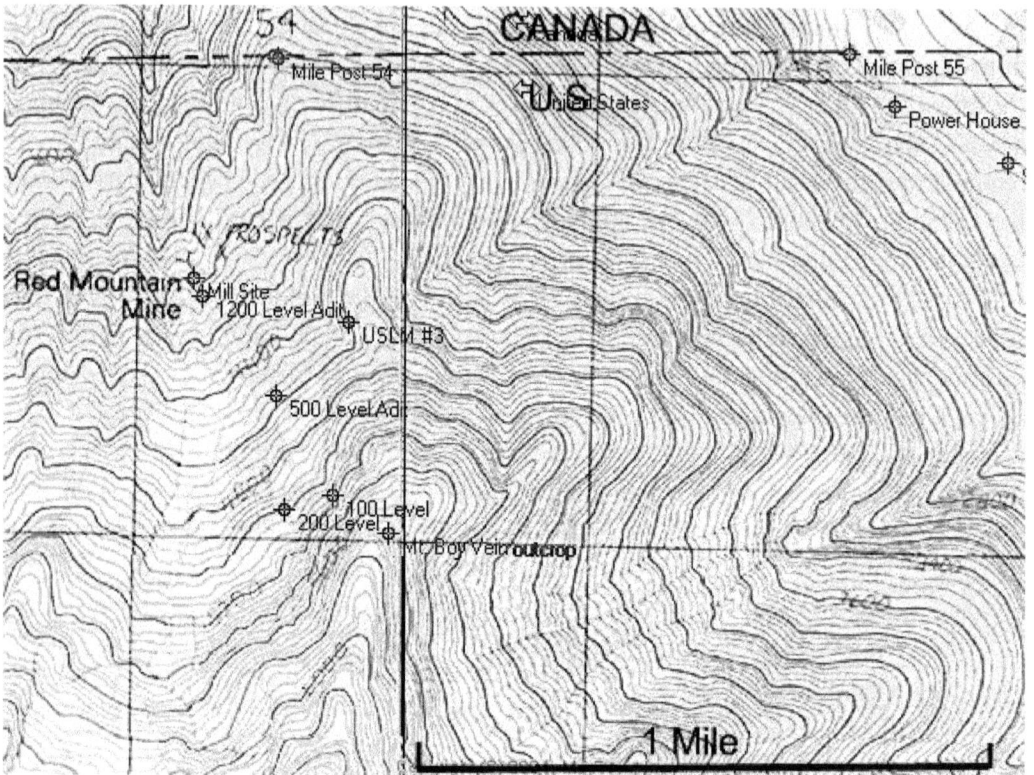

Figure 83. Topographic map of the Boundary Red Mt. mine showing several levels of mine entrances, the mill site, and power house on Silesia Creek. (USGS topographic map)

The mineralogy, origin of the gold, and other data indicate that the gold at the Boundary Red Mt. mine and the Lone Jack mine are essentially identical except for the host rocks. Gold at the Boundary Red Mt. mine occurs in five unconnected quartz veins in the Yellow Aster schist and fine-grained metamorphic diorite (a rock much like granite except that it lacks quartz). Like the mineralization at the Lone Jack mine, the gold-bearing veins formed during two stages: (1) fractures in the rock were filled with quartz from percolating solutions, and (2) later, shearing from crustal movements produced grinding up of the quartz in the veins, allowing hot, gold-bearing solutions to infiltrate the earlier-formed quartz and deposit gold. The principal ore mineral is native gold with minor amounts of silver. The ground up quartz fissures contain distinctive brown, wavy bands of iron oxide in the quartz. Most of the gold is so fine-grained that it cannot be seen without magnification.

Figure 84. Outcrop and mine entrance (arrow) of one of the quartz veins at the Boundary Red mine. (Photo from Wolff, et al., 2008)

Five veins were discovered at the Boundary Red Mt. site, but all of the extracted gold has come from one vein known as the Red Mt. vein, which can be traced for more than 1,500 feet. It strikes N14°E, with dips ranging from 58° to 70° southeast. Veins are exposed on the surface for as much as 900 feet, but are cut by faults so veins are usually not continuous for more than about 100 feet. Some of the veins are truncated by faults, some die out in zones of ground up quartz, and some pinch out in the host rock. Thickness of the veins varies considerably, averaging about three feet but changing from a thin smear to 10 feet in a few feet laterally along the vein.

Mining occurred by entry from four levels totaling more than 7,000 feet in length and reached about 950 feet below the surface. Including mined out areas and secondary tunnels, the total mined length was probably more than ~10,000 feet. The mill processed about 80,000 tons of ore.

**Figure 85. Portal at the 500 foot level of the Boundary Red mine.
(From Wolff et al., 2008)**

Figure 86. Mill at the Boundary Red Mt. mine about 1930. (Photo from Wolff et al., 2008)

Figure 87. Boundary Red Mt. mill about 1930. (Photo from Wolff et al., 2008)

Great Excelsior Mine

In 1900, W. H. Norton and others discovered gold along Wells Creek above Nooksack Falls (Fig. 88) about six miles east of the town of Glacier. Two years later, Norton created the Great Excelsior Mining Co. and constructed a 20–stamp mill on the mine site.

The occurrence of gold at the Great Excelsior mine is different than at the Lone Jack and Boundary Red Mt. mines. It occurs in breccia, rock composed of angular fragments of rock, in a zone about 400 feet long and 270 feet wide dipping 55° west. It is underlain by fractured slate. The host rocks are part of the Wells Creek volcanic group, a 4,000-foot thick sequence of volcanic breccia with interbedded marine shale, siltstone, sandstone, and volcanic tuff (ash). Rocks above the slate have been altered by pervasive, hot, gold-bearing solutions.

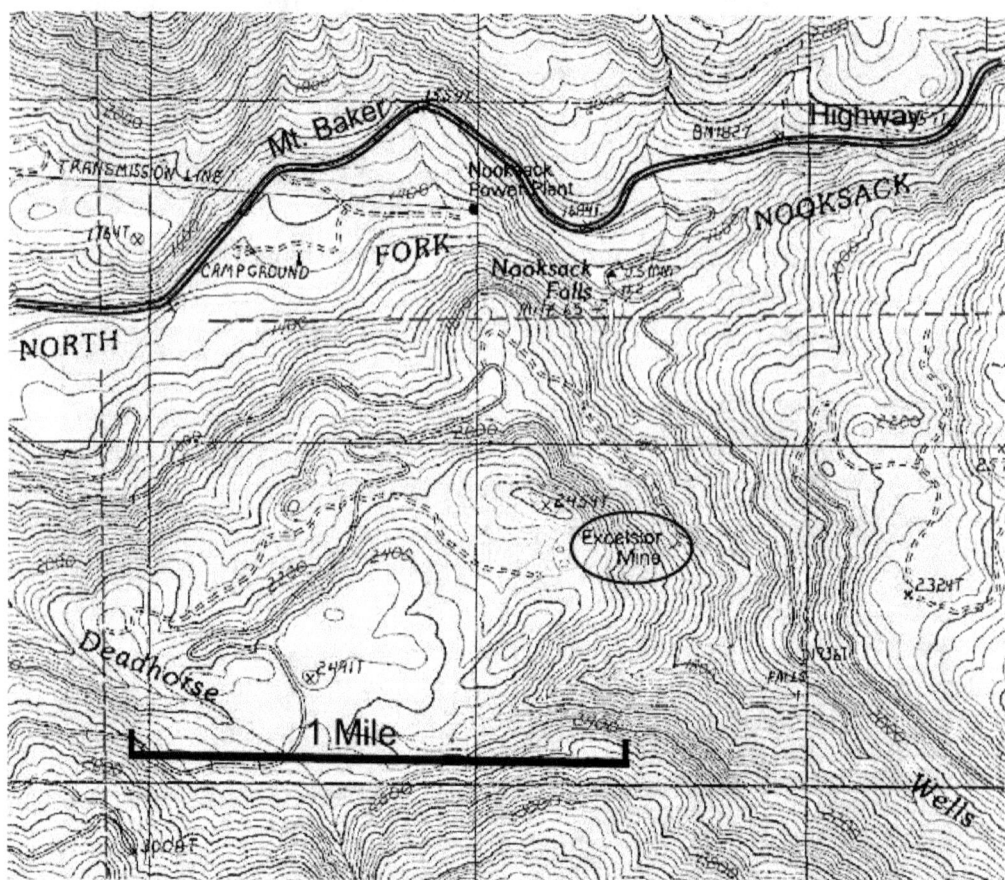

Figure 88. Topographic map of the Great Excelsior mine site. (USGS topographic map)

Figure 89. Great Excelsior Mill Level mine portal. (Photo from Wolff et al., 2004)

In 1902, a 20-stamp mill was built on steep slopes immediately below the Mill Level tunnel. However, during milling operations to recover the gold, about half of the finely disseminated gold was lost. The mill was rebuilt in 1914 to improve gold recovery using a cyanide system, but that also proved unsatisfactory and the mill was shut down in 1917.

Renewed exploration took place between 1972 and 1993, largely as a result of changes in the economics of mining large tonnages of low grade ore and large swings in the price of gold and silver. In 1980, the market price of silver peaked at a historic high of $48 per ounce, only to return to $5 per ounce eighteen months later. In 1983, gold rose to $600 per ounce, but fell to about $300 per ounce a short time later. About 10,000 tons of ore was mined and netted only about $69,000.

CHAPTER 2

GEOLOGIC STRUCTURES—FOLDING AND FAULTING OF THE ROCKS

SUMAS GRABEN—A MAJOR DOWNDROPPED BLOCK BETWEEN VEDDER MT. AND SUMAS FAULTS

Northwest Washington lies within the structurally complex continental margin. Collision of the Pacific plate against the North American and Juan de Fuca plates provides strong driving forces that cause faulting and rock deformation. As the oceanic Juan de Fuca plate is overridden by the westward-moving North American plate, northwest Washington is subject to three distinct types of earthquake activity: (1) earthquakes within the continental crust, (2) deeper earthquakes within the oceanic plate, and (3) very large earthquakes at the boundary between plates. Most earthquakes in NW Washington are fairly shallow and occur within the continental crust of the North American plate. The Sumas graben is a structural block that has dropped down between the Vedder Mt. and Sumas faults (Figs. 90-92) and filled with unconsolidated sediments.

Vedder Mt. fault

The Vedder Mt. fault makes a prominent, linear, NE-SW–trending escarpment that truncates bedrock structures along the sides of the Sumas Valley (Figs. 90–91). This has been known since 1904, but new data now indicates that both the Vedder Mt. and Sumas faults are larger and more active than previously thought. The Vedder Mt. fault extends from British Columbia into Washington near Sumas, continues southwesterly across Whatcom County, and appears to extend westward to Sucia Island in the San Juan Islands. The fault is at least 65 miles long and may be even longer. The Sumas fault parallels the Vedder Mt. fault and also extends southwesterly from British Columbia across Whatcom County.

Wells have penetrated 1,000 feet of unconsolidated sediment in the down–dropped block (Fig. 92). The thickness of unconsolidated sediments indicates the amount of down–dropping of the valley floor and shows that it is geologically young. The total offset along the Vedder Mt. fault is at least 1,000 feet plus 1,500 feet for the height of the fault scarp along the side of Vedder Mt.. Thus, the total amount of movement on the fault is apparently at least 2,500 feet. The surface expression of the fault disappears beneath unconsolidated glacial deposits that cover western Whatcom County.

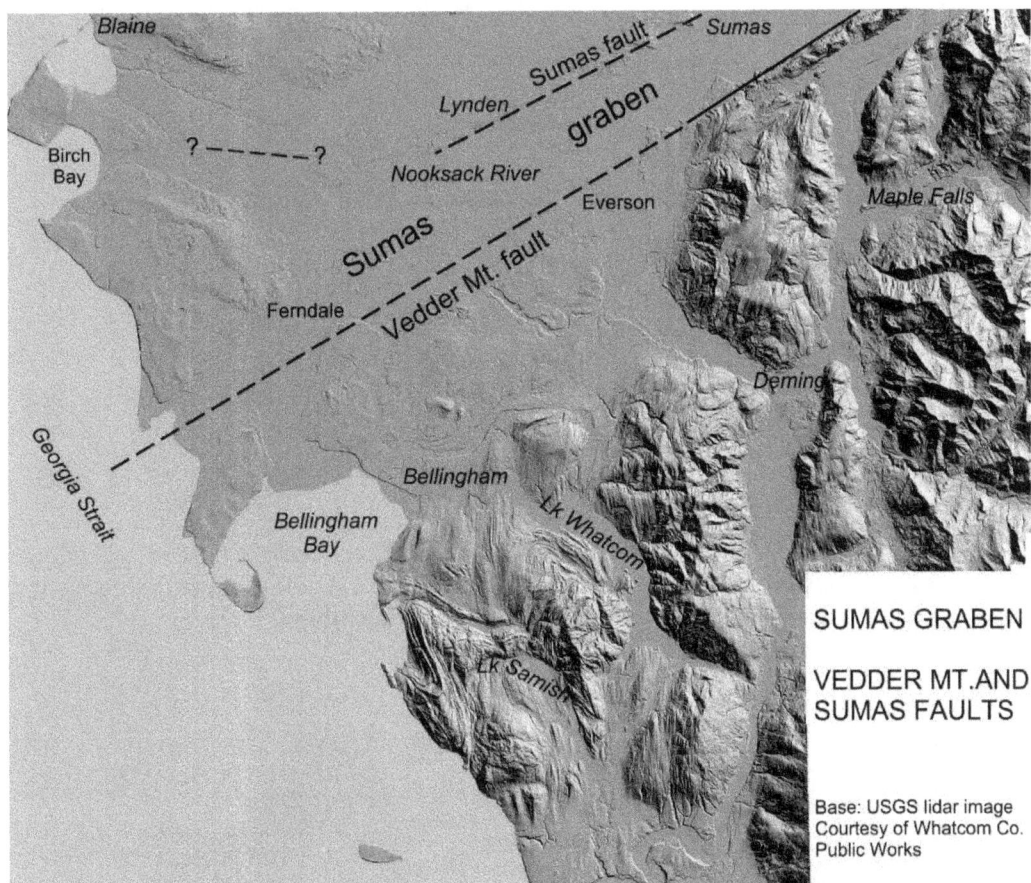

Figure 90. Major faults in Whatcom County. The Sumas graben, Veddar Mt. and Sumas faults. (USGS lidar image provided by Whatcom Co.)

Geologic structures below sea level at Sucia Island in the San Juan Islands are truncated by a prominent, linear fault that may be the continuation of the Vedder Mt. fault. Deformation of sea floor sediments east of Sucia Island is consistent with eastward continuation of the fault toward the mainland where the fault is covered on land by late Pleistocene glacial deposits. Eastward extrapolation of this fault lines up reasonably well with the western extension of the Vedder Mt. fault and suggests a possible correlation.

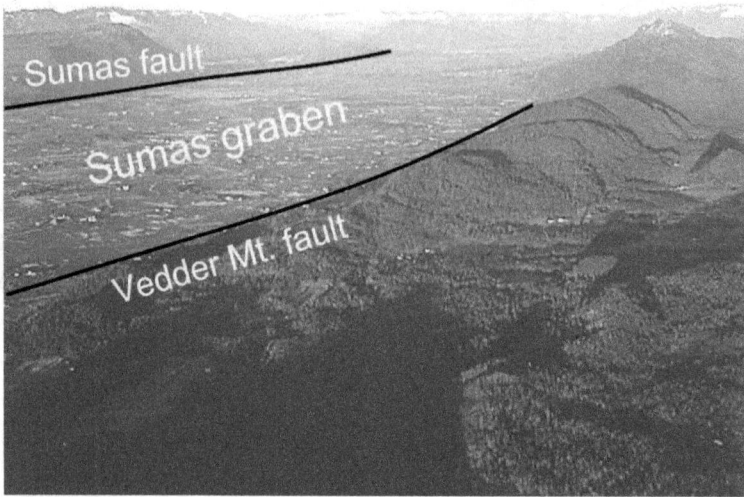

Figure 91. Sumas graben, bounded by the Vedder Mt. and Sumas faults. View looking north into Canada from the international border.

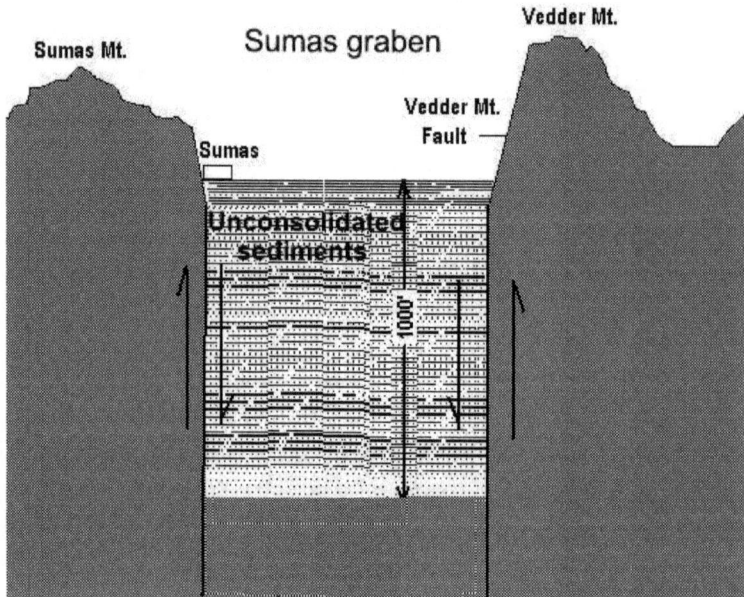

Figure 92. Geologic cross section of the Sumas graben, which is a down-dropped block between the Vedder Mt. and Sumas faults.

The Sumas fault

The Sumas fault parallels the Vedder Mt. fault and extends southwesterly from British Columbia through Sumas and across Whatcom County (Figs. 90-92). The trace of the fault across Canadian Sumas Mt. is quite obvious, but to the southwest, it is covered by thick Ice Age sediments and is difficult to locate in the subsurface. Because of the deep fill of the Sumas graben and fault geometry, the offset along the Sumas fault must be fairly similar to that along the Vedder Mt. fault. What is less certain is the exact location of the Sumas fault beneath the glacial sediments southwest of Sumas.

Between Blaine and Lynden, an anomalous escarpment (Fig. 93) may be a fault scarp representing the surface offset of movement along the Sumas fault. The scarp is more or less in line with the projected westward extension of the Sumas fault, but more importantly, the scarp exhibits features that are incompatible with an erosional origin: (1) The scarp is separated from the modern Nooksack River by a high area so the scarp was not eroded by the river, (2) it is anomalously straight—scarps made by rivers are notable for irregular cusps and curves, (3) the south (lower) side of the scarp is not a flat fluvial surface as it would be if the scarp had been eroded by a stream, and (4) the eastern end of the scarp feathers out into the Pleistocene outwash plain to the east with no indication of stream activity on the south side. All of these features taken together strongly suggest that the scarp was not made by stream or marine erosion, leaving fault offset as the most likely cause for the scarp.

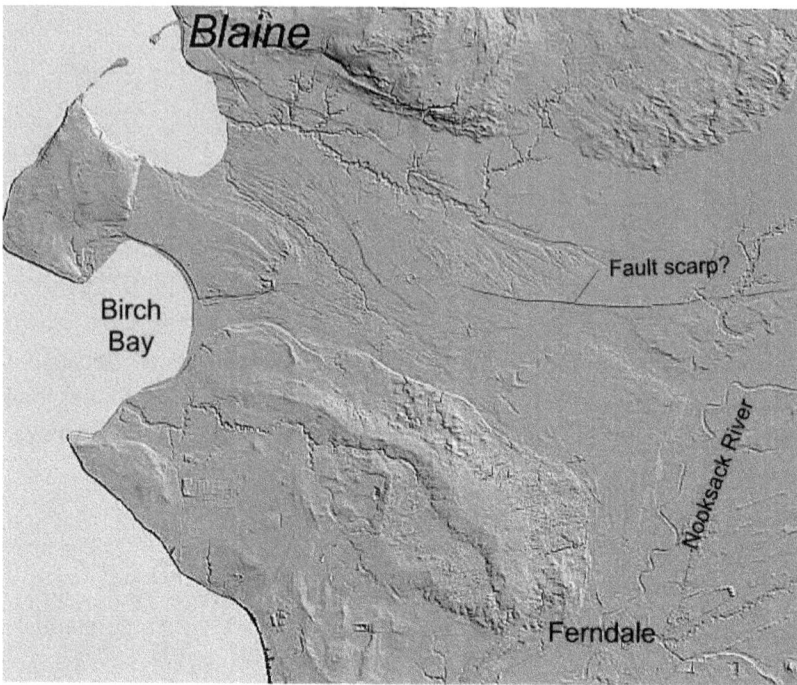

Figure 93. Lidar image of an escarpment near Custer that appears to be offsetting of the land surface by a fault scarp. (USGS lidar image provided by Whatcom Co.)

EARTHQUAKES—ON SHAKY GROUND

Earthquakes have been recorded along the traces of the Sumas and Vedder Mt. faults since 1964, indicating that the faults are presently active (Fig. 94). A magnitude 5.0 earthquake occurred along the projected extension of the Sumas fault in 1964 and a magnitude 6.0 earthquake occurred in 1909 in the San Juan Islands near the western end of the Vedder Mt. fault.

Earthquake records show a remarkable pattern in this part of NW Washington. About 1,800 shallow earthquakes have been recorded since 1969 within the region. Eleven historic earthquakes with magnitudes from 4.0 to 7.4 occurred within 50 km of Whatcom County. These historic events include the largest recorded Washington earthquake of magnitude 7.4 in 1872.

Analysis of earthquake epicenters on the Vedder Mt. fault indicates NE–SW fault movement upthrown on the south. In addition to the recent seismic activity associated with the Vedder Mt. and Sumas faults, large numbers of earthquakes have been recorded in the Deming area (Fig. 94). The region near Deming is among the most active earthquake zones in the state with hundreds of quakes recorded since 1969, including the

98

April 14, 1990 quake (Richter magnitude 5.2), which was one of the five largest quakes in the Pacific Northwest between 1965 and 1992. Hundreds of foreshocks and thousands of aftershocks bracketed the main event.

Although earthquake epicenters line up along the Vedder Mt. and Sumas faults, the reason for the clustering of earthquakes in the Deming area is not clear. The earthquakes near Deming do not show a linear trend typical of movements along a fault, but rather scatter within a circular area. In general, earthquakes in the Deming area since 1969 have released 10 times more energy in the shallow crust than the surrounding area.

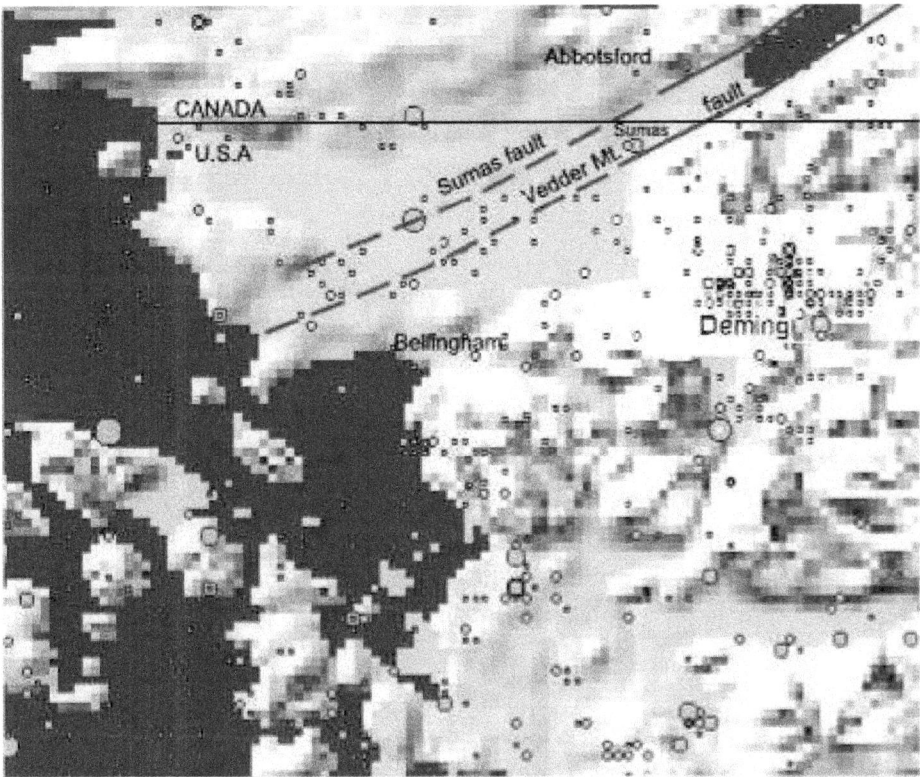

Figure 94. Earthquake epicenters in Whatcom County since 1969. Each dot represents a recorded earthquake. The size of the dots is proportional to the magnitude of the earthquake.

The April 14th, 1990 magnitude 5.2 Deming earthquake was the largest shallow earthquake (9 miles) in NW Washington in recent years. It has been interpreted to have occurred on a shallow (less than 2.5 miles) low angle fault. Hundreds of foreshocks and thousands of aftershocks bracketed the main event. A magnitude 4.8 earthquake occurred before the main shock.

The 1990 Deming earthquake is significant when assessing potential earthquake hazards in Whatcom County because (1) it is one of the five largest earthquakes to have occurred in the Pacific NW between 1965 and 1992, (2) it was the largest known low angle thrust fault earthquake in Washington, (3) it was the largest shallow earthquake in Washington or Oregon since at least 1920, and, perhaps most important, (4) the Deming earthquakes show a principal crustal stress direction unique to the rest of the region.

POTENTIAL EARTHQUAKE HAZARDS

Seismic shaking

The intensity of an earthquake and its potential for damage depend on several factors:

(1) The larger the size of an earthquake, the greater the intensity of shaking. Bigger earthquakes cause more damage.
(2) Nearness to the epicenter of the quake.
(3) The nature of the material beneath the ground. For example, the size of seismic waves is much greater in clay than in bedrock.
(4) The type of construction—large structures are more vulnerable than smaller ones and structures that vibrate with the same wave frequencies as those of earthquake waves may undergo greatly amplified shaking.

The town of Sumas is especially vulnerable to earthquake damage because (1) it lies directly over the Sumas fault, (2) the Vedder Mt. fault lies only 2–3 miles away on the south side of the valley, (3) the thick fill of clay and silt in the valley (1000 ft) amplifies earthquake waves, and (4) lake clay and silt beneath the valley floor are subject to possible liquefaction.

Bellingham is only a few miles south of the Vedder Mt. fault as crosses the county (Fig. 90). This is a major fault with a lot of displacement and a record of recent activity so it must be considered when assessing potential earthquake hazards.

Ground failure, liquefaction

During earthquakes, failure of the ground beneath structures can be highly destructive. Seismic waves can cause clay, silt, and fine sand to act like liquids so that the ground literally flows, a process known as liquefaction. The floor of Sumas valley is filled with thick layers of unconsolidated lake clay and silt that lie on more than 1000 feet of other fine–grained sediment. These sediments are vulnerable to shaking that could cause liquefaction.

CHAPTER 3

MT. BAKER

Mt. Baker is the fourth-highest mountain in Washington at 10,778 feet and is one of five major volcanic cones along the crest of the Cascade Range in Washington. It is part of a long chain of volcanoes extending from northern California to British Columbia along the Cascade crest. All of the volcanoes are similar in composition, presumably related to the same fundamental crustal activity, and have been constructed during the past several hundred thousand years by eruption of lava and ash.

Figure 95. Mt. Baker from over Bellingham. The snowy peaks to the right are the Twin Sisters and Lake Whatcom lies in the center foreground.

Indigenous Indian natives called the peak Koma Kulshan, meaning white sentinel. In 1790, a Spanish expedition under Manuel Quimper sailed from Nootka on Vancouver Island to explore the Strait of Juan de Fuca. During that six-week voyage, Gonzalo Lopez de Haro drew detailed charts and a sketch of the mountain, which the Spanish named "La Gran Montana del Carmelo." In 1792, Captain George Vancouver explored the area and, while anchored in Dungeness Bay on the south shore of the Strait of Juan de Fuca, he renamed the mountain for 3rd Lieutenant Joseph Baker of HMS Discovery.

Mt. Baker has three principal eruptive centers (Figs. 96, 97): (1) the summit crater covered by an ice and snow dome, (2) Sherman Crater immediately south of the summit, and (3) remnants of an older cone, the Black Buttes (Figs. 96 98, 99), now deeply eroded. In addition, a small cinder cone and an associated lava flow occur at Schreibers Meadow on the south flank of the mountain.

Figure 96. Mt. Baker summit cone (left) and the Black Buttes (right).

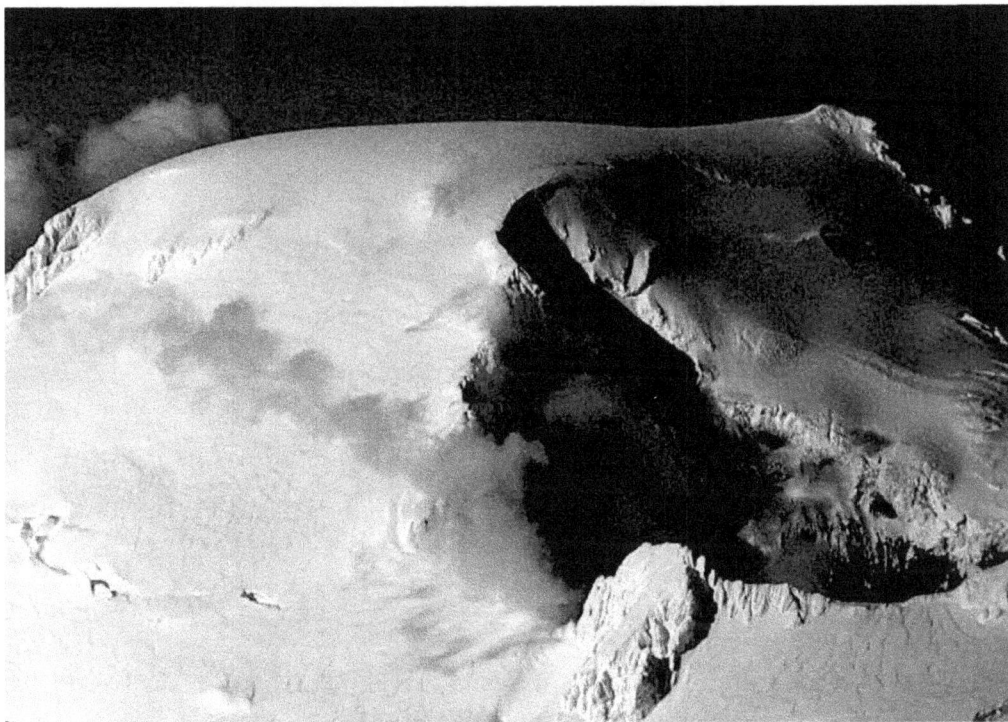

**Figure 97. Mt. Baker summit cone (left), 10,778 feet above sea level
and Sherman Crater (right)**

THE BLACK BUTTES—AN ANCIENT
MT. BAKER VOLCANO

The present summit cone of Mt. Baker was built upon the eroded wreckage of an older volcano whose remnants are now exposed as the Black Buttes on the western flank of the main cone (Figs. 98, 99). The Black Buttes consist of two peaks, Lincoln Peak (9,096 ft.) (Fig. 100) and Colfax Peak (9,443 ft.), which have been deeply dissected by erosion of glaciers on their slopes. Remnants of lava flows exposed in the rock walls of the Black Buttes dip in opposite directions away from a former summit cone now long eroded (Figs. 98, 99). The relative degree of erosion, the overlap of Mt. Baker volcanic rocks upon those of the Black Buttes, and the topographic position of older flows from the Black Buttes volcano indicate the relative antiquity of the Buttes.

103

Figure 98. Mt. Baker summit cone (left) and the eroded flanks of the older Black Buttes volcanic cone (right).

Figure 99. Colfax Peak (left) and Lincoln Peak (right), eroded remnants of the Black Buttes volcanic cone.

Figure 100. Lincoln Peak, the western Black Butte. Lava flows and volcanic deposits of volcanic fragments are inclined 25-30° to the west (right) on the sides of the breached Black Buttes volcano.

Flows in the vicinity of Colfax Peak dip eastward toward the present summit of the main cone of Mt. Baker, whereas flows near Lincoln Peak dip in the opposite direction, indicating that the peaks are remnants of a volcanic cone whose central vent was between the two peaks. Similar degrees of erosion characterize lava flows that cap ridges high above adjacent stream valleys between the Buttes and the Nooksack River.

North of the Black Buttes, most of lower apron of the cone has been eroded away and to the east, remnants disappear beneath the younger cone of Mt. Baker. Farther from the Black Buttes, stacks of lava flows hundreds of feet thick cap ridges high above the modern streams.

Ten isotope dates of Black Buttes lavas yield ages of 330,000 to 350,000 years. Three isotope dates from the topmost Black Buttes lava flows have been dated between 288,000 and 306,000 years and probably represent the final activity of the Black Buttes volcano.

HIGH, LAVA-CAPPED RIDGES THAT WERE FORMERLY VALLEYS

At a number of places, lava flows, probably from the Black Buttes volcano, went down old stream valleys and now cap ridges high above modern streams. Such inversion of topography is common in the area between the Black Buttes and the North Fork of the Nooksack River.

Table Mt., a high, flat-topped ridge west of Artists Point, is a remnant of lava flows that have become topographically inverted by deep erosion. The summit of Table Mt. is about 1,500 feet above Bagley Lakes. The headwall of a deep glacial basin (cirque) at its northeast edge drops about 1,200 feet from the Table Mt. summit to the cirque floor at Bagley Lakes. The lava flow capping the ridge at Table Mt. is 300-400 feet thick.

Samples of the lava collected from Table Mt. were originally dated in the 1970s by isotope analysis at 400,000 ± 100,000 years. A more recent isotope date by the U.S. Geological Survey gave an age of 309,000 ± 13,000 years.

Figure 101. Table Mt., a ridge capped by lava that formerly occupied a valley 309,000 years ago at Heather Meadows.

Figure 102. Table Mt. (the white ridge) looking west. The linear ridge of Table Mt. marks the floor of a former stream valley whose sides have been eroded away.

Figure 103. Table Mt. lava flow looking east. Chain Lakes in the foreground. The sinuous pattern of the ridge reflects the former stream valley down which the lava flowed.

The prolonged erosion necessary to develop topographic inversion, from valley floors down which the lava flowed to the present erosional remnants on a ridge crest, indicates that a great deal of erosion has taken place since the lava flow was emplaced. The amount of erosion to topographically invert the Table Mt. flows to a ridge crest is about 2,000 feet.

The inverted topography at Table Mt. remains as testimony that a great deal of erosion has occurred during the past 300,000 years. The original valley sides, probably composed of ashy material having much less resistance to erosion than the lava flows in the valleys, eroded much faster than the lava, resulting in inversion of topography so that the flow originally occupying a valley, now makes up a resistant ridge.

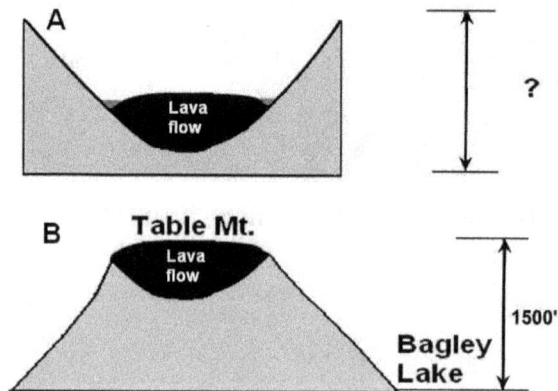

Figure 104. Topographic inversion of Table Mt. lava flow. (A) Emplacement of Table Mt. lava. (B) Erosion of less resistant valley sides, leaving the lava flow as a ridge crest.

The lava flows that floor the Heather Meadows area below Table Mt. are well exposed in road cuts along the highway leading to the ski area (Figs. 105, 106). A lava flow along the highway below Heather Meadows has been isotope dated at 301,000 ± 8,000 years.

Many of the lava flows at Heather Meadows exhibit columnar jointing (rock fractures) (Figs. 105, 106). Columnar jointing consists of six-sided fractures that are caused by contraction of the lava during cooling. As lava cools, it shrinks and forms six-sided fractures that pull away from one another, forming columns. The long axis of each column forms at right angles (vertical) to the surface of the lava flow (the cooling surface), so as long as the flow of the lava is uniform, the columns will be upright. However, if the cooling surface is disturbed, as when lava flows over irregular topography, the columns will become distorted (Fig. 107).

Figure 105. Columnar jointing in a lava flow from Mt. Baker along the highway below Heather Meadows.

Figure 106. Columnar jointing, Mt. Baker lava flow below Heather Meadows.

Figure 107. Swirling columnar jointing in a lava flow along the road from Austin Pass to Artists Point. The columns are distorted because the lava flowed over a steep slope that disturbed the normal temperature gradient within the flow, which controls the formation of columns.

A number of other flow remnants have similar ages and are considered to be part of the Black Buttes eruptive phase. Coleman Pinnacle south of Table Mt. along Ptarmigan Ridge consists of lava, isotope dated at 305,000 ± 6,000 years. Lava flows about 300 feet thick that cap ridge crests at Cathedral Crag and Baker Pass on the south flank of Mt. Baker have compositions similar to other Black Butte lavas and have been isotope dated at 331,000 ± 18,000 years and 333,000 ± 12,000 years respectively. Remnants of other flows that once extended several miles down the Glacier Creek area cap Bastille Ridge and in Smith Basin north of the cone have compositions similar to Table Mt. and have been isotope dated at 322,000 ± 9,000 and 322,000 ± 12,000 respectively.

A cliff seven miles east of Nooksack Falls, visible from the Mt. Baker Highway across the Nooksack River, is a ridge–capping lava that flowed down an ancient valley but now stands as a mesa about half a mile long. The lava flow is about 300 feet thick, with prominent columnar jointing (fractures) that formed during cooling and contraction of the lava. The top of the flow is about 700 feet above the floor of the modern canyon and its ridge-capping topographic position suggests a possible correlation with the Table

Mt. flows, but the isotope age of the flow is only 202,000 ± 9,000 years, 100,000 years younger than the Table Mt. flows. The rate of erosion needed to invert the valley-filling lava flow to what is now a ridge crest is remarkable.

Near the junction of Wells Creek and the Nooksack River, a 1,300 foot long, 250 foot thick lava flow remnant caps a ridge now 800 feet above river level, indicating that more than 800 feet of erosion has occurred since the flow was emplaced. An isotope date from this flow in 1975 was 400,000 ± 100,000 years. However, an isotope age of 149,000 ± 5,000 years was obtained more recently. The deep erosion to leave the lava flow capping a ridge 800 feet above the river suggests correlation with other ridge capping flows of about 300,000 years so the reason for the discrepancy in ages is uncertain.

A third flow remnant near Nooksack Falls occurs 650 feet above the floor of Wells Creek near the abandoned Excelsior Mine and has been isotope dated at 114,000 ± 9,000.

MT. BAKER SUMMIT CONE AND SHERMAN CRATER

The main summit cone of Mt. Baker (Grant Peak) (Figs. 108-109) is about 1,300 feet in diameter but its crater is not conspicuous because it is filled with ice and snow and visible only late in the summer. Radio echo sounding indicates that the ice in the summit crater is at least 300 feet thick. The ice flows out of the carter to the north at the head of the Roosevelt glacier.

Most of the summit cone has been built of lava flows and interbedded fragments ejected in the past 40,000 years, although isotope dates of old flows near the base of the cone are as old as 140,000 years. The youngest lava flows are 9-20,000 years old.

Sherman crater is just south of the main summit cone and is about 2000 feet in diameter (Fig. 109). Unlike the main summit crater, which is filled with ice and snow, Sherman crater makes a deep pit about 1,500 feet below the main summit. The crater is younger than the main summit crater and eruptions from it seem to have been entirely ash and pumice that have mantled the surrounding area.

Figure 108. Mt. Baker summit cone (Grant Peak) from the north.

Figure 109. Sherman crater and Grant Peak, Mt. Baker.

VOLCANIC ASH

Five volcanic ashes have been recognized in meadow exposures around Mt. Baker–
Schreibers Meadow scoria, Mazama ash, Rocky Creek ash, Cathedral Crag ash, and the 1843
Mt. Baker ash (Fig. 110).

Volcanic unit	Radiocarbon age

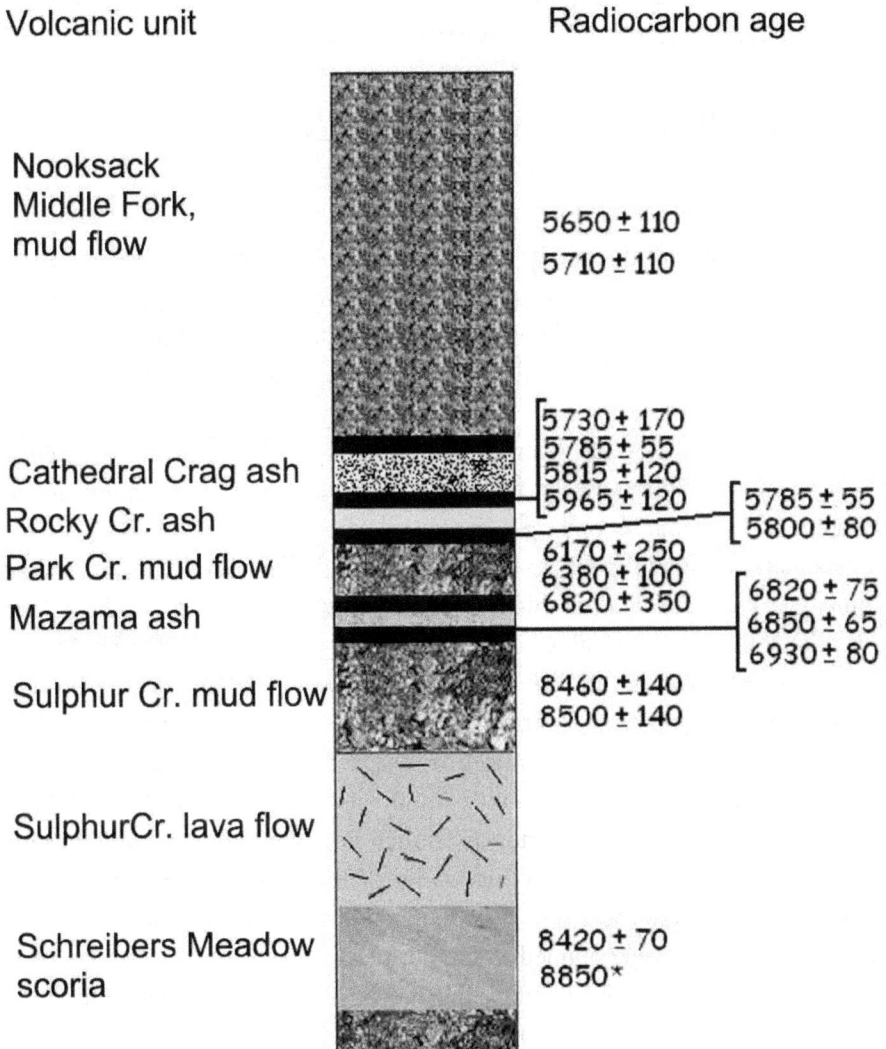

Nooksack Middle Fork, mud flow — 5650 ± 110, 5710 ± 110

Cathedral Crag ash — 5730 ± 170, 5785 ± 55, 5815 ± 120, 5965 ± 120 [5785 ± 55, 5800 ± 80]

Rocky Cr. ash

Park Cr. mud flow — 6170 ± 250, 6380 ± 100, 6820 ± 350 [6820 ± 75, 6850 ± 65, 6930 ± 80]

Mazama ash

Sulphur Cr. mud flow — 8460 ± 140, 8500 ± 140

SulphurCr. lava flow

Schreibers Meadow scoria — 8420 ± 70, 8850^*

**Figure 110. Sequence of volcanic ashes, mudflows, and lava flows
around the flanks of Mt. Baker.**

113

The oldest of three ash layers in the Boulder Creek valley on the south flank of Mt. Baker lies beneath two lava flows, two other ash layers, and other volcanic deposits. Its thickness and particle size suggest that it originated from Mt. Baker. A small wood fragment beneath the ash was dated at 8,700 ± 1,000 years, but the date is tentative because of the limited amount of sample material available. A younger ash, separated from the one described above by a lava flow, is less than an inch thick and consists of very fine particles, suggesting a distant source. A still younger fine-grained ash layer in Boulder Valley, overlain by a lava flow, is ~3 inches thick.

SCHREIBERS MEADOW CINDER CONE, SCORIA, AND LAVA FLOW

Lava, emanating from a cinder cone at Schreibers Meadow (Figs. 111–113) on the south flank of Mt. Baker, flowed at least seven miles down Sulphur Creek (Fig. 111) about 8,500 years ago. The cinder cone lies along the Sulphur Creek fault, which extends up the Nooksack Middle Fork valley, crosses the divide into the Sulphur Creek valley, and continues to the east.

Drainage in the valley has been displaced by the lava flow, and Sulphur Creek and Rocky Creek each flow between the valley sides and the flow, leading to the unusual situation of two streams occupying a single valley.

A distinctive, coarse–grained, orange–brown scoria (lava with many holes from escaping gas bubbles), was erupted from the cinder cone in Schreibers Meadow and mantles the southwestern flank of Mt. Baker in a northeasterly fallout plume (Figs. 114, 115). The scoria is up to six feet thick near the vent but thins to the northeast where meadows that have not been overrun by glaciers show a distinctive orange color. The scoria thins to three feet about a mile from the vent and is only about 2–4 inches thick five miles northeast of the cone. Pumice fragments are up to an inch in diameter near the cone, but decrease to about a quarter of an inch near the end of the fallout plume. At its type locality in an alpine meadow just south of the Squak Glacier, the scoria contains abundant charcoal, which was radiocarbon dated at 8,420 ± 70 radiocarbon years old. An age of 8,800 has also been obtained at another place.

Downvalley from the cinder cone, the scoria is overlain by the Sulphur Creek lava flow and a volcanic mud flow containing wood radiocarbon dated at 8,500 radiocarbon years old.

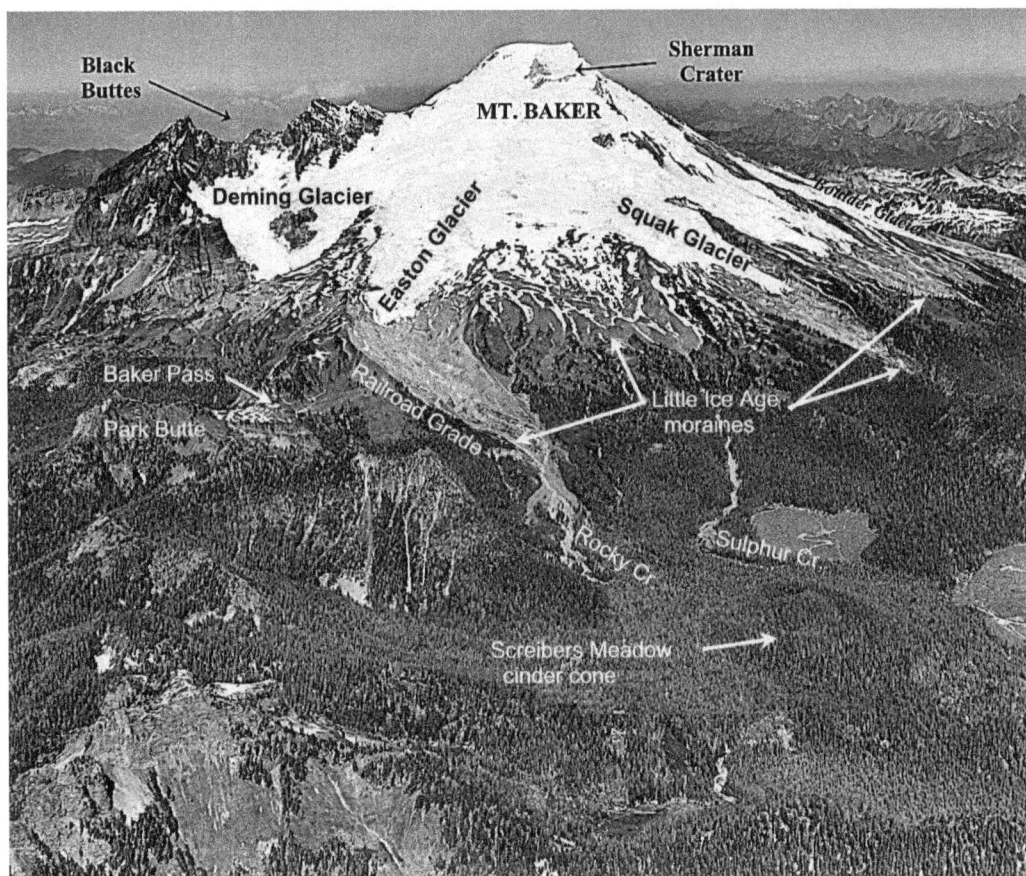

Figure 111 Schreibers Meadow cinder cone and other localities on the south flank of Mt. Baker.

Figure 112. Schreibers Meadow cinder cone.

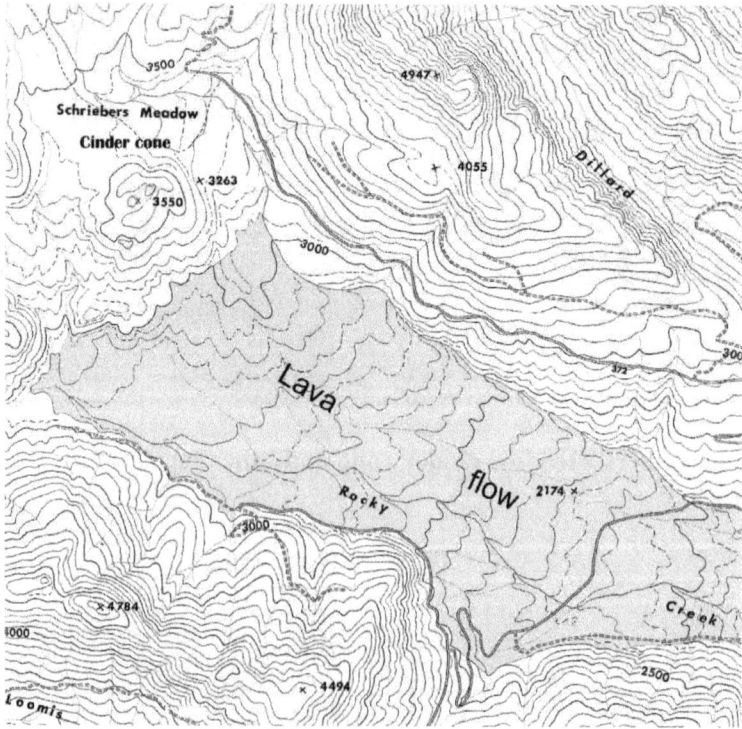

Figure 113. Schreibers Meadow cinder cone and lava flow.

116

Figure 114. Schreibers Meadow scoria at Schreibers Meadow.

Figure 115. Schreibers Meadow scoria below the Squak glacier on the SW
flank of Mt. Baker.

Mazama Ash

6,850 radiocarbon years ago, Mt. Mazama in southern Oregon erupted violently, covering the entire Pacific Northwest with ash. Collapse of the volcano following the last eruption created Crater Lake. In the North Cascades, Mazama ash consists of conspicuous, fine-grained, yellow-orange ash, generally 1-3 inches thick, composed mostly of volcanic glass with abundant pumice. The ash has a distinctly different chemical composition than ashes from Mt. Baker and can be readily identified. Radiocarbon dating of peat above and below Mazama ash at seven localities indicates that the ash is 6,850 radiocarbon years old. Mazama ash is visible in almost every exposure along trails and creek banks around Mt. Baker.

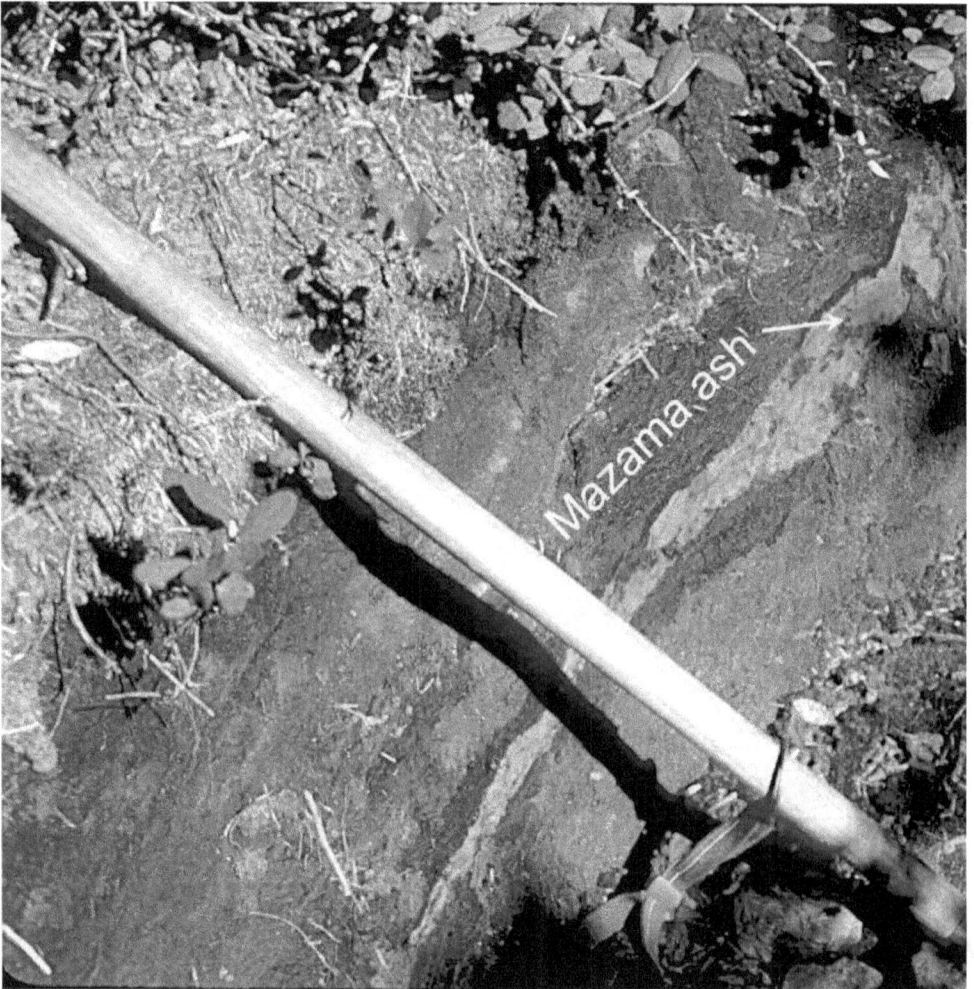

Figure 116. Mazama ash at Heather Meadows.

Rocky Creek Ash

A light-gray, sandy ash overlying Mazama ash in meadows around Mt. Baker (Fig. 117) was first dated near Heather Meadows in 1975. It was later named the Rocky Creek ash and its type locality designated on the SW flank of Mt. Baker (Figs. 119, 120) where it is underlain by peat and Mazama ash and overlain by Cathedral Crag black sandy ash. Rocky Creek ash consists mostly of well-sorted, sand-silt–size rock particles with a few glass shards. In most places, the ash is a few inches thick, but in some places it reaches thicknesses of 12 inches.

Figure 117.   Mazama, Rocky Creek, and Cathedral Crag ashes at Heather Meadows.

Microprobe analyses show that the composition of glass shards in the Rocky Creek ash is distinctly different from glass in Mazama ash but quite similar to the overlying Cathedral Crag ash from Mt. Baker.

The age of the Rocky Creek ash is tightly constrained at close to 5,800 radiocarbon years by dates of 5,785 ± 55, and 5,800 ± 80 radiocarbon years below the ash and by dates of 5,730 ± 170, 5,815 ± 120, and 5,965 ± 120 radiocarbon years above the ash. Rocky Creek ash is present in many road and trail cuts throughout Heather Meadows, Artists Point, Ptarmigan Ridge, and Chain Lakes, as well as many other meadows around Mt. Baker.

119

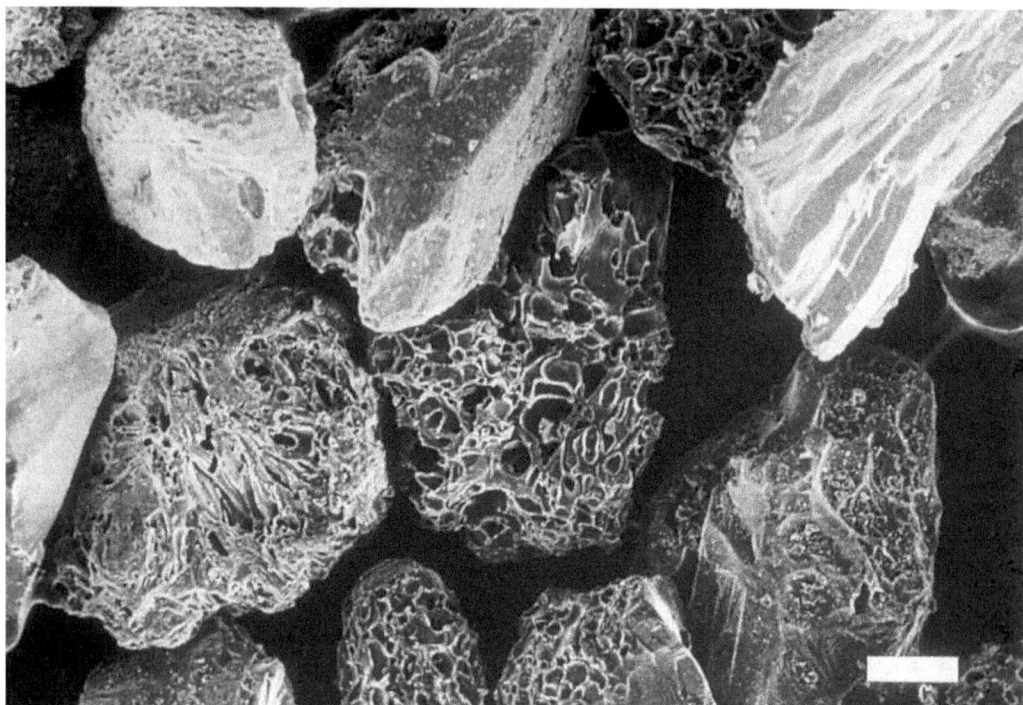

Figure 118. Microscopic view of typical pumice (dark particles with many gas bubbles) and glass (white particles with no gas bubbles) in volcanic ash.

Cathedral Crag Ash

The Cathedral Crag ash was defined and dated at its type locality near Cathedral Crag on the SW flank of Mt. Baker above Schreibers Meadow (Fig. 119). It consists of massive, black, sandy ash and is the thickest and most widely distributed ash erupted from Mt. Baker. At Heather Meadows, it is up to 30 inches thick. The prominent peninsula that extends into Picture Lake from the south consists of Cathedral Crag ash and Mazama ash.

The Cathedral Crag ash consists of well sorted, crystal-rich, coarse sand-sized particles. Microprobe composition of the glass resembles other Mt. Baker ashes, but is quite distinct from Mazama ash.

Ages of 5,785 ± 55, and 5,800 ± 80 radiocarbon years ago have been obtained from organic material beneath Cathedral Crag ash, and a date of 5,780 years was obtained from plant fragments at the base of the ash in upper Swift Creek between Artists Point and Mt. Shuksan. These dates are very close to the 5,800 radiocarbon year age of the underlying Rocky Creek ash.

Figure 119. Mazama ash and type locality of the Rocky Creek and Cathedral Crag ashes along the trail above Schreibers Meadow.

Because Cathedral Crag ash commonly lies directly on Rocky Creek ash with no intervening meadow peat or sediment, the Rocky Creek ash was probably a steam eruption of mostly rock particles ripped from the sides of the vent, closely followed by eruption of the Cathedral Crag ash containing glass from erupting lava that cooled in the air.

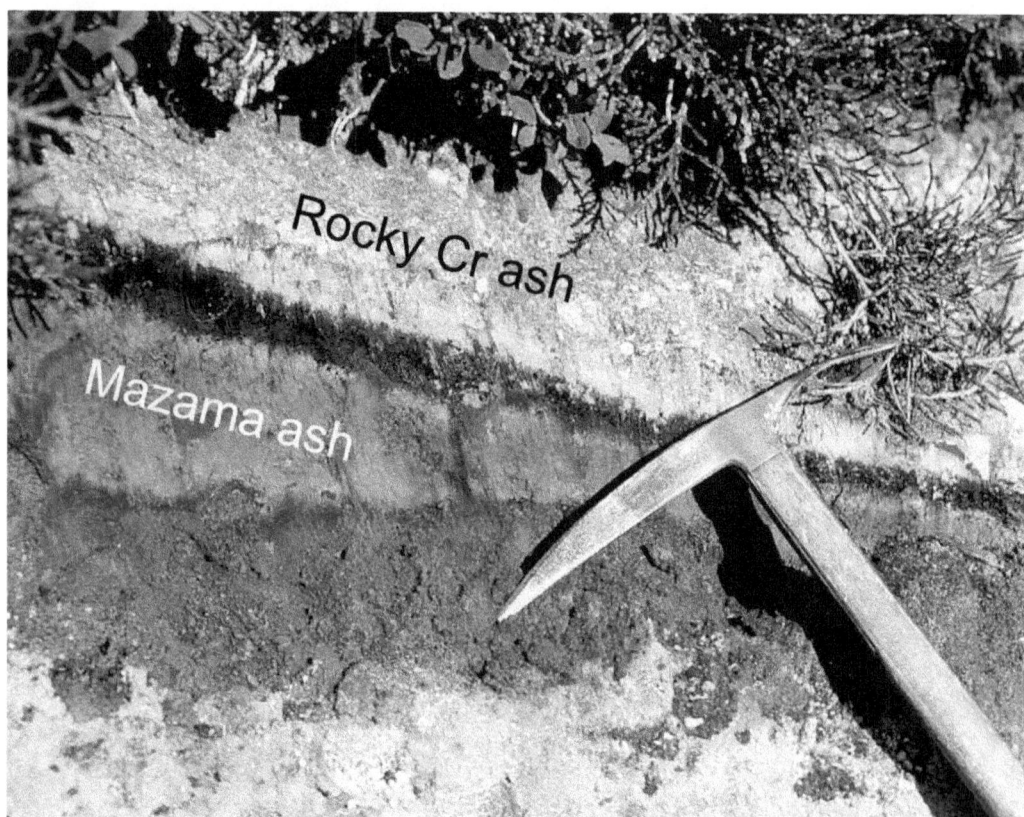

Figure 120. Mazama and Rocky Creek ashes above Schreibers Meadow.

1843 ash eruption

A volcanic deposit composed of gray–white silt sized ash and rock fragments up to five inches in diameter occurs in meadows on the south flank Mt. Baker. Most of the larger rock fragments have been chemically altered and are coated with sulphur. It was most likely erupted from Sherman Crater by a steam eruption of Mt. Baker in 1843 that ripped fragments from the sides of the vent. No lava flows were associated with the eruption.

HISTORIC ERUPTIONS OF MT. BAKER

When Captain Vancouver sailed into Puget Sound in April, 1792, he observed a high volcano that was emitting "smoke" (actually steam), and he named the peak Mt. Baker, after Lieutenant Baker of his crew.

In June 1792, the Spanish expedition of Dionisio Alcalá Galiano and Cayetano Valdés, while anchored in Bellingham Bay, reported:

"During the night, we constantly saw light to the south and east of the mountain of Carmelo [Mt. Baker] and even at times some bursts of flame, signs which left no doubt that there are volcanoes with strong eruptions in those mountains."

Figure 121. Steam from Sherman Crater.

In 1843, explorers reported widespread ash that fell "like a snowfall" and that the forest was "on fire for miles around." Indians reported that nearby rivers were clogged with ash and many salmon perished. In 1854, George Davidson of the Coast and Geodetic Survey noted "the summit of this mountain obscured by vast rolling masses of dense smoke, which in a few minutes reached an estimated height of two thousand feet above the summit, and soon enveloped it entirely."

In 1858, steam activity was noted by several people and on November 26, 1860, passengers traveling by steamer from New Westminster to Victoria reported that Mount Baker was "puffing out large volumes of smoke, which upon breaking, rolled down the snow-covered sides of the mountain, forming a pleasing effect of light and shade."

Figure 122. Vigorous steam jets from the west rim of Sherman Crater, 1975.

The first ascent of Mt. Baker by the E. T. Coleman party in 1868 revealed emission of steam from several fumaroles in Sherman Crater. On July 8, 1891, a climbing party reported, "The opening was fifty by seventy-five feet in circular shape and puffs of smoke issuing from the interior would vanish in the light air at a short distance." On September 9, 1891, another climbing party observed that the crater" . . . "is filled with snow, except in the center, where there is a circular opening from which steam and sulphurous vapors are constantly escaping. In fact we noticed the presence of the sulphurous vapors 2000 feet below the summit."

A photograph of the crater taken from the summit shortly after 1891 shows a circular steam vent with steam coming from it. In 1900, John A. Lee described 'smoke' rolling up from the crater in great clouds and a large round vent in the snow from which steam and sulfur fumes were issuing. In 1906, C. F. Easton reported that "vapors of sulphur are continually emitted and steam jets issue from myriads of vents with great violence." Since then, the crater has been seen by numerous climbing parties and observed from airplanes, but published reports of activity are scarce. Air photographs taken in 1947, 1950, and 1955 show some steam activity or glacial melting, and mild steam began again in the early 1960s.

Except for a small steam vent on the north side of Mt. Baker, present steam activity is restricted to Sherman Crater, a circular depression about 1,500 feet in diameter, breached at the east rim by a notch leading to the Boulder Glacier. The floor of the crater lies about 1,300 feet below the summit ice dome and is covered with glacial ice and snow.

On March 10, 1975, a tall column of steam was observed rising from Sherman Crater. The steam, clearly visible from the lowland west of the mountain, rose to altitudes of about 1,500 feet above the crater floor. Early reports by local residents of dark clouds coming from the crater were later confirmed by observation of much dark material on the crater floor and on the upper part of Boulder Glacier. Although mild steam activity had been observed in the Sherman Crater area for many years, association with ejection of solid material was unusual, and the height and vigor of the steam jets was considerably greater than had been observed since the 1850s.

Steam activity in Sherman Crater on March 11, 1975, consisted of a large jet, forcibly ejecting steam under pressure to heights of 300 to 800 feet above the crater floor, and five other vents were emitting steam. The second largest emission was from a circular vent in the glacier, estimated to be about 100 feet in diameter near the north wall of the crater.

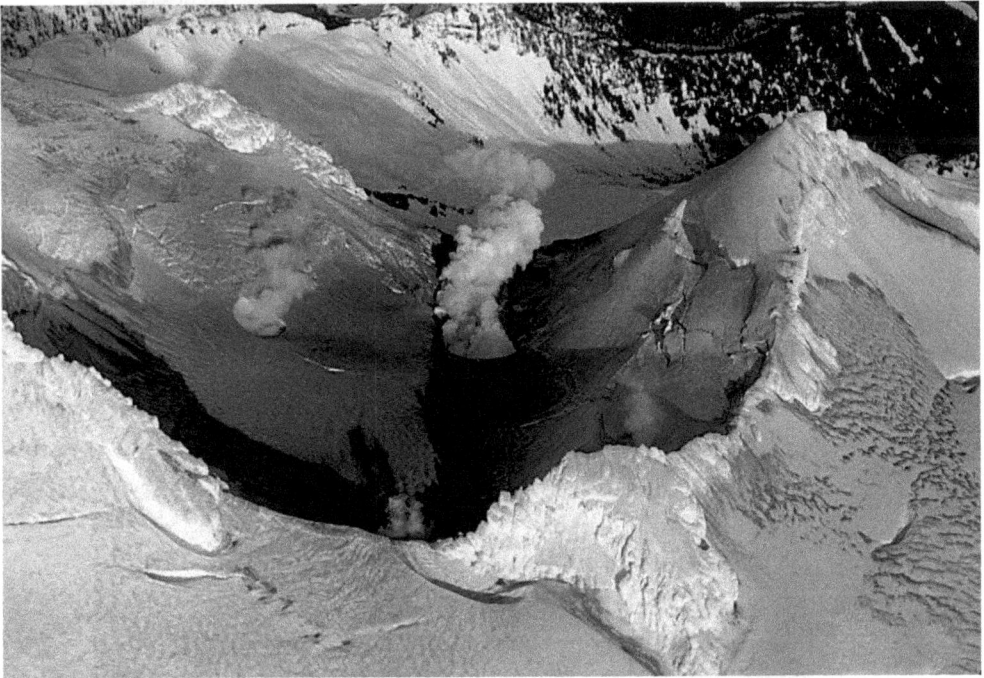

Figure 123. Ash–coated floor of Sherman crater from steam eruption, March 11, 1975. The ash was composed of old rock ripped off the walls of vents by steam.

On April 1, 1975, a circular depression about 200 feet in diameter appeared in the glacier just south of the center of the crater floor (Figs. 124). Further collapse was observed on April 10, followed by complete collapse of an ice roof into a hole about 150 feet deep in the glacier (Figs. 125–127).

125

Figure 124. Circular collapse of the glacier on the floor of Sherman crater.

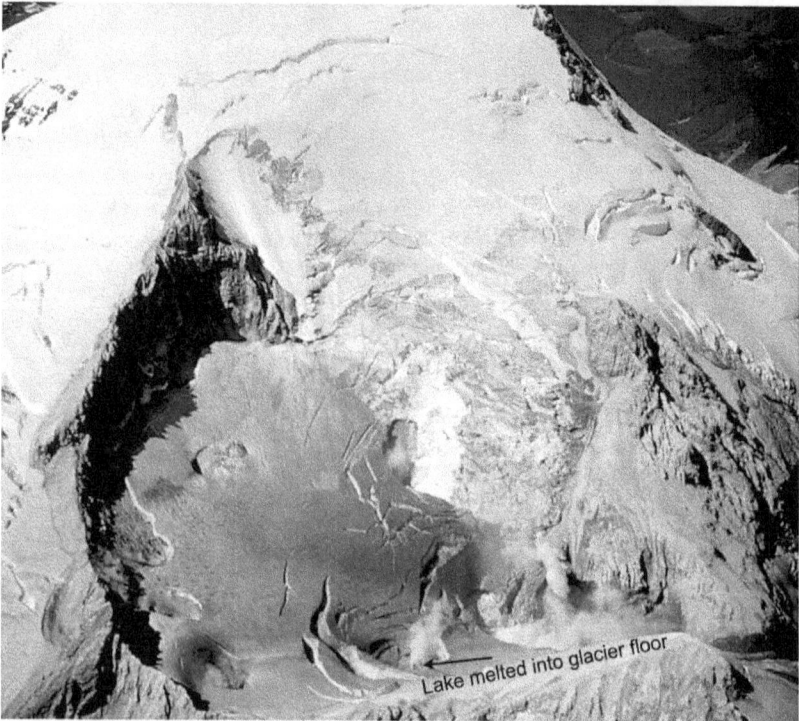

Lake melted into glacier floor

Figure 125. Beginning of collapse of the glacier on the floor of Sherman Crater to form a circular depression, April 10, 1975.

Figure 126. Large hole melted in the glacier on the floor of Sherman crater. The dark color on the glacier is rock debris blown out of steam vents in the crater.

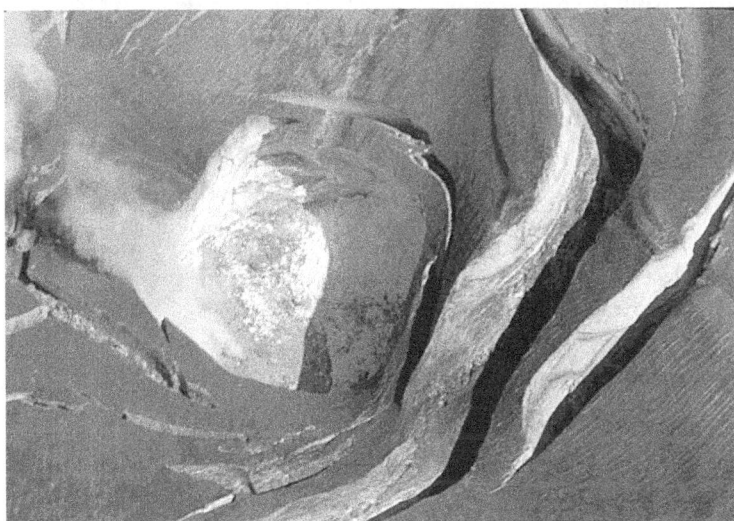

Figure 127. Lake forming in a hole melted into the glacier on the floor of Sherman Crater. The dark surface of the glacier is ash blown out of steam vents.

In early April, 1975, the glacier at the north wall of Sherman Crater began to develop large crevasses as ice pulled away from the rest of the glacier and fell into the vent below. The vent began to enlarge rapidly by collapse of glacial ice early in May. A large elongate pit about 300 feet deep developed at the base of the north wall, causing the glacier above to break up (Fig. 128).

Figure 128. Breaking up of the glacier on the floor of Sherman Crater at the north wall of Sherman Crater.

Figure 129. Steam activity in Sherman Crater, 1980.

128

The largest steam activity observed was on June 29, 1975 when a column of steam about 2,500 feet high rose from Sherman Crater. By the end of July, the glacier on the crater floor was severely crevassed as ice continued to collapse generally toward the east rim and locally toward the vent areas. Large concentric crevasses extend nearly across the entire crater, and adjacent to the north vent, the glacier was cut by intersecting crevasses that isolated huge blocks of ice (Fig. 128). Soon after 1975, steam activity in the crater subsided to a lower level where it has since remained.

Figure 130. Big steam eruption from Sherman crater on Mt. Baker, June 29, 1975.

MT. BAKER GLACIERS—NATURE'S THERMOMETERS

Glaciers are masses of ice and granular snow formed by compaction and recrystallization of snow, lying largely or wholly on land and showing evidence of past or present movement. Permanent snowfields that persist through the summer melt season are not considered glaciers because they do not move. Glacial ice begins as light, fluffy snow. Snow that survives the warm summer season is progressively transformed into granules by compaction, pressure melting, and refreezing into glacial ice.

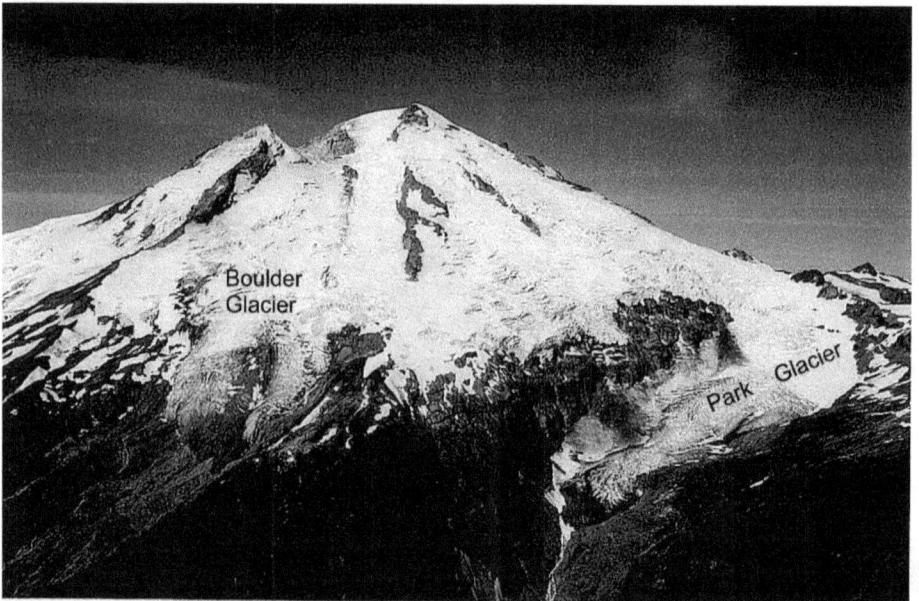

Figure 131. Boulder and Park glaciers on the south flank of Mt. Baker.
(Photo by Austin Post)

Glaciers are fed largely by winter snow, and ice is lost mostly by summer melting. Accumulation of snow and ice is greater than summer melting in the upper part of a glacier, and melting of ice is greater than accumulation of winter snow in the lower part of a glacier. Thus, we would expect to see the lower part of glaciers melt away each summer, but they don't because the ice lost to the glacier in the lower zone is replaced by movement of ice from the upper accumulation zone. Thus, the position of the terminus of a glacier is determined by the relative amount of winter accumulation and summer melting and the rate at which ice flows from the upper accumulation zone to the lower melting zone. If snow accumulation exceeds melting over a number of years, the amount of glacier ice will increase and the glacier terminus will advance. If summer melting exceeds winter snowfall over time, the glacier terminus will recede.

Thus, glaciers are nature's thermometers. They fluctuate back and forth, much like mercury in a thermometer. When the climate warms, glaciers recede; when the climate cools, glaciers advance. Although glaciers are also affected by precipitation (snowfall), summer temperatures strongly affect glacier melting and fluctuations of glacier termini. Because glaciers persist for thousands of years and leave a record of their former margins, they are very useful for determining climate changes in the past. Mt. Baker glaciers have been especially useful in reconstructing past climate changes because they have been especially sensitive to climate changes. Glaciers are to climate what mercury is to thermometers.

Mt. Baker has more glaciers than any other mountain in Washington except Mt. Rainier. It is one of the snowiest places on Earth, holding the world record for snowfall in a single season, 1,140 inches (95 feet) in 1999.

THE LITTLE ICE AGE (1300 A.D. TO THE 20TH CENTURY)

The Medieval Warm Period (900 AD to 1300 AD) was a time when global temperatures were slightly warmer than at present and human populations prospered. The Vikings were able to colonize Greenland, wine could be grown in northerly latitudes, food was relatively plentiful, and civilization thrived. At the end of this warm period, about 1300 AD, temperatures suddenly dropped several degrees in about 20 years at high latitudes, and the cold period that followed, known as the Little Ice Age, lasted for several centuries. The cold climate was devastating to many regions, especially Europe and other areas at mid to high latitudes. The population of Europe had become dependent on cereal grains as a food supply during the Medieval Warm Period, and when the colder climate, early snows, shorter growing season, violent storms, and recurrent flooding swept Europe, massive crop failures occurred repeatedly. Crops that had fed populations for generations suddenly failed, season after season. About one third of the population of Europe perished from famine and disease. Winters were bitterly cold in many parts of the world and glaciers worldwide advanced. The advance of glaciers in the European Alps in the mid–17th century encroached on farms and villages. Glaciers in Greenland and elsewhere advanced and pack ice extended southward in the North Atlantic. Sea ice surrounding Iceland extended for miles in every direction, closing many harbors, and the population of Iceland decreased by half. The Thames River in London froze over, and Viking colonies in Greenland died out because they could no longer grow enough food. In parts of China, warm weather crops that had been grown for centuries were abandoned. In North America, early settlers experienced exceptionally severe winters. In Whatcom County, all of the glaciers on Mt. Baker advanced and left a record of glacial deposits far downvalley.

Global temperatures have risen about 1° F per century since the Little Ice Age, but the warming has not been continuous. Numerous 25-35 year warm/cool cycles appear in the glacial and ocean records, as well as in isotope records in Greenland ice cores.

Coleman Glacier

The Coleman glacier is named after Edmund Thomas Coleman, an Englishman from Victoria, B.C., who, with Edward Eldridge, John Tennant, David Ogilvy, and Thomas Stratton made the first ascent of Mt. Baker on August 17, 1868.

The Coleman glacier is the largest glacier on Mt. Baker. It originates at the saddle between the main summit cone and the Black Buttes and flows down the north flank of the mountain to the headwaters of Glacier Creek (Fig. 132). The glacier terminus fluctuates back and forth as the climate changes, each time leaving a ridge of debris at the ice margin (moraine) to mark where it has been, or, in the case of recent changes in the ice margin, leaving a trim line in the adjacent forest where the ice stripped away the trees.

Figure 132. Coleman (right) and Roosevelt (left) glaciers
(Photo by Austin Post, 1962)

During the Little Ice Age (1300 AD to about 1915 AD), the Coleman glacier extended much farther downvalley than it does at present (Fig. 133) and left multiple moraines and trimlines in the adjacent forest, marking its former position.

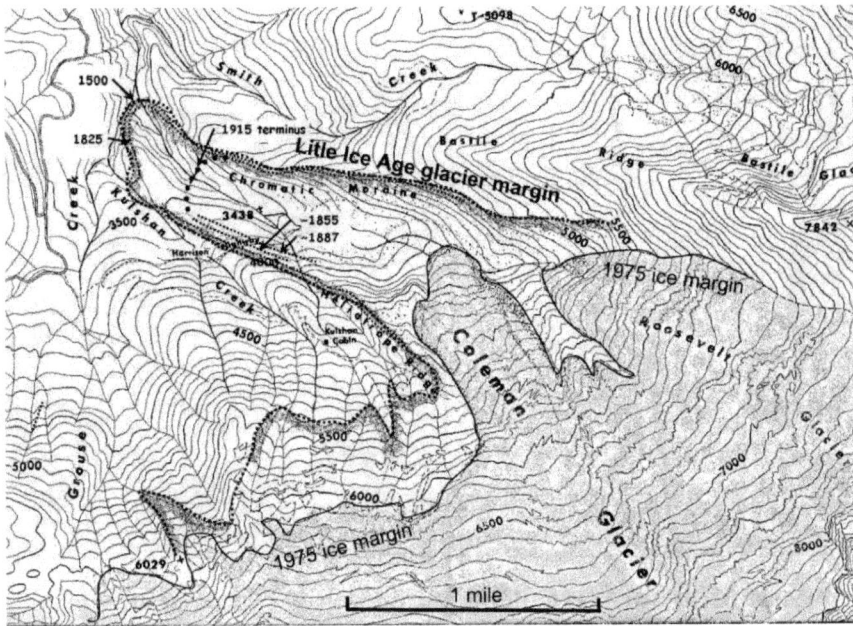

Figure 133. Former ice extent of the Coleman glacier dating back to the Little Ice Age in about 1500 AD. (Modified from USGS topographic map)

The oldest Little Ice Age moraine has trees growing on it dating back to the 1500's. A buried forest on the Coleman glacier moraine (Fig. 134), radiocarbon dated at 680 ± 80 and 740 ± 80 radiocarbon years ago, grew during the Medieval Warm Period atop an older moraine. The forest was buried by a Little Ice Age moraine when the glacier overtopped the forest. This buried forest is exposed in the lateral moraine adjacent to the Coleman glacier and may be seen by walking to the moraine from the old Kulshan cabin site (upper Kulshan Creek, Fig. 133) and walking down the crest of Heliotrope Ridge.

Annual rings from trees growing on successively younger moraines show moraine–building episodes in the 1600s, ~1750, ~1790, ~1850, and ~1890. Each of these moraines represents a cold climate episode. The Little Ice Age glacial advance of the 1600s occurred in a cold period that took place during the Maunder Solar Minimum, a time when the sun had almost no sun spots. The 1790 glacial advance occurred during another cool period from 1790 to 1820, known as the Dalton Solar Minimum when the sun again had very few sunspots. The 1890–1910 cool period, when many of the cold temperature records in North America were set, also occurred during a time of few sun spots.

Figure 134. Medieval Warm Period forest buried by lateral moraine of the Coleman glacier during the Little Ice Age.

During the cold period from 1890 to about 1910, the Coleman glacier extended well downvalley from its present terminus, almost all the way down to the 1500 AD Little Ice Age ice margin (Figs. 133, 135). When the climate warmed after 1915, the glacier receded upvalley about a mile where it fluctuated during the rest of the century.

Ice margins of Mt. Baker glaciers since 1940 are shown on air photos taken every several years in a remarkable collection by glacial geologist Austin Post. The 1940 positions of the Coleman and adjacent Roosevelt glacier margins are shown on Figure 137. The glaciers retreated upvalley to their 1947 position (Fig. 137) during the 1915 to 1945 warm period. The global climate then shifted from warm to cool, and the Coleman glacier advanced vigorously (~2,500 feet) until 1979 (Fig. 137). The glacier margin extended well downvalley from its 1940 position and the Coleman and Roosevelt glaciers nearly merged (Fig. 137). In 1977, the climate suddenly shifted back to warm and the glaciers once again retreated upvalley. The 2005 terminus of the glacier was about 1500 feet upvalley from its 1979 position.

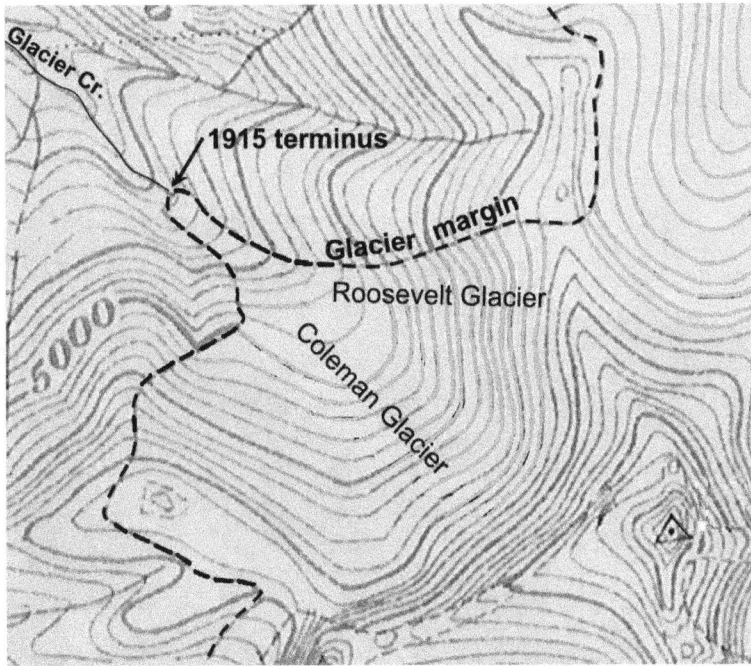

Figure 135. Map of the Coleman glacier in 1915. Note that the Coleman and Roosevelt glaciers were merged into a single terminus. (Modified from USGS map)

Figure 136. The Coleman glacier early in the last century (about 1915). The Coleman and Roosevelt glaciers merged into a single terminus with a ridge of debris between them. Compare these glaciers with those in Fig. 132 (1962).

These glacier fluctuations closely follow the global cooling record and indicate that the ~30 years warming and cooling cycles seen in the glacial record mimic global climate and ocean temperature changes. Thus, these historic glacial fluctuations record global climate changes.

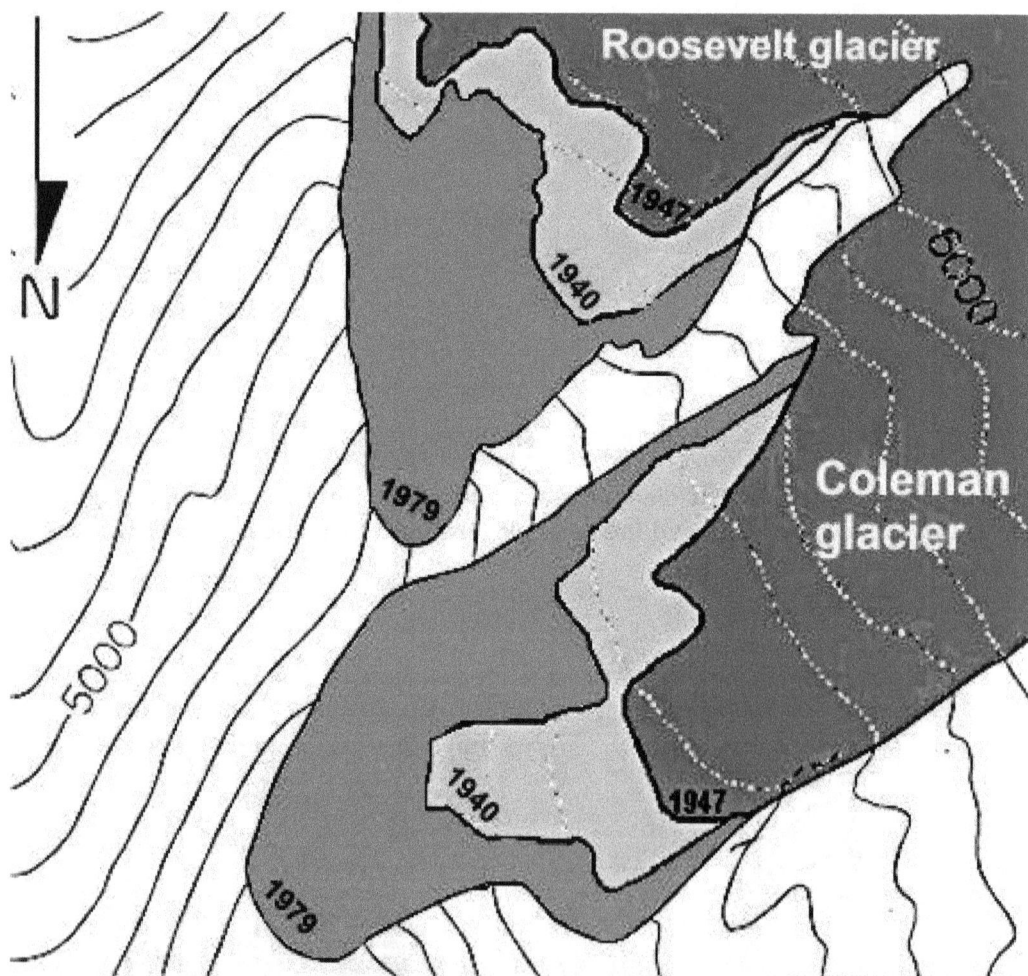

Figure 137. Historic fluctuations in the Coleman and Roosevelt glaciers. In 1915, the glacier terminus was more than a mile downvalley and retreated to the 1947 position during the ~1915 to ~1945 warm period. From 1947 to 1979, the terminus advanced downvalley beyond the 1940 position but not as far as the 1915 position. (Modified from Harper, 1992)

Deming Glacier

The Deming glacier starts high on the main summit cone, flows into a deep basin on the Black Buttes, then out of the basin into the lower valley (Fig. 138). Avalanches from the Black Buttes fall onto the glacier, and the north side of the lower glacier is usually covered with rock debris. The glacier occupies a deep trough scoured out by ice when the glacier was much larger than it is now.

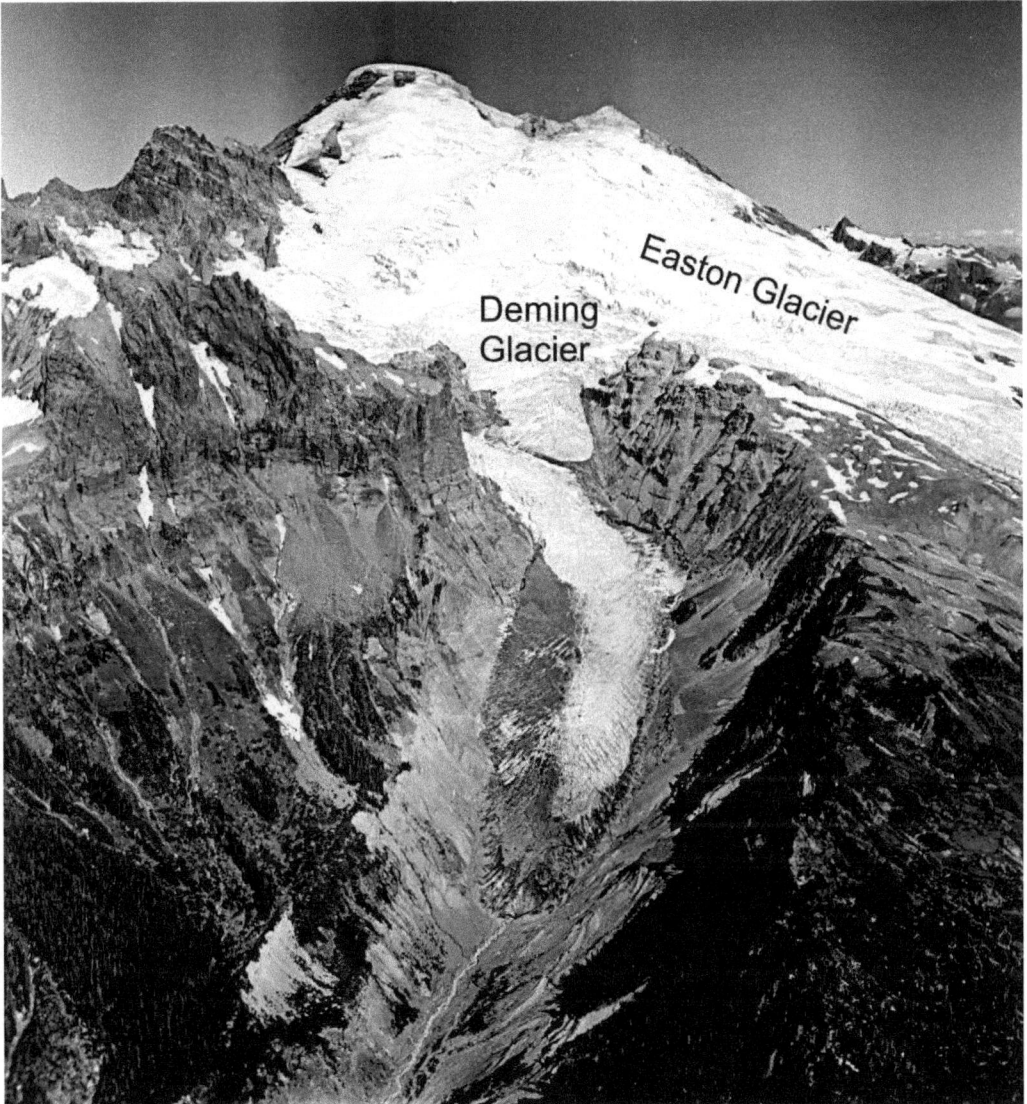

Figure 138. Deming glacier in the 1960s. (Photo by Austin Post)

137

Of all of the Mt. Baker glaciers, the Deming glacier has the longest record of glacial advances and retreats. At the lower end of the valley near Ridley Creek, logs buried in glacial deposits have been radiocarbon dated at ~10,500 ^{14}Cyears old (Fig. 139). During this period of time, known as the Younger Dryas, the Cordilleran Ice Sheet was still present in the lowland north of Bellingham and glaciers all over the world were vigorously advancing.

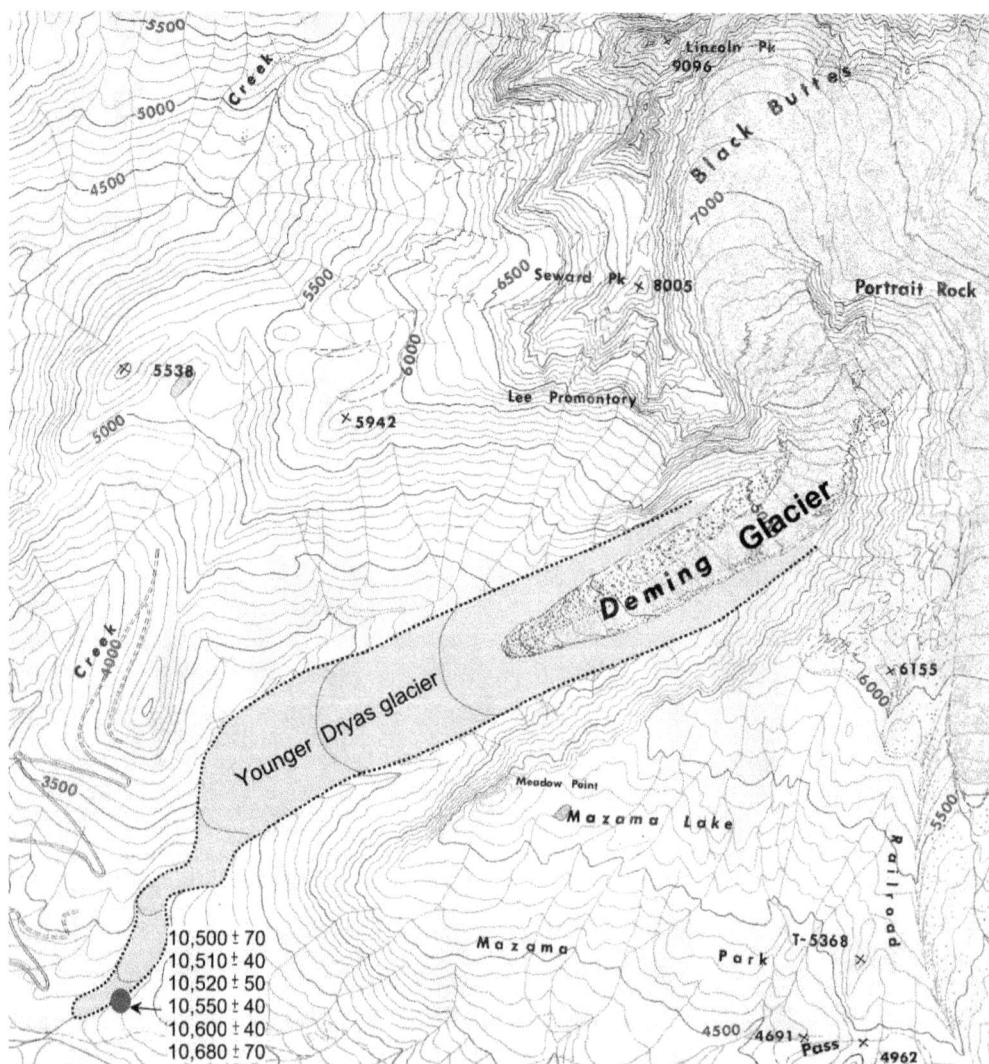

Figure 139. Reconstructon of the Deming glacier during the Younger Dryas cold period at the end of the last Ice Age. Six radiocarbon dates from logs buried in glacial deposits in the lower valley of the Deming glacier near Ridley Creek indicate an age of ~10,500 ^{14}C years.

138

Logs and buried forests also occur farther upvalley at several localities. Logs at two localities between the present glacier terminus and Ridley Creek, near the confluence with the Nooksack Middle Fork, have been radiocarbon dated at 2,960 and 2970 years old and 2,440 and 2205 years old (Figs. 140, 141). These buried logs imply that the Deming glacier extended at least that far downvalley at the time of the dates. Access to the sites of the buried logs is difficult because of lack of a trail, a deep gorge at the lower end of the valley, and growth of closely spaced alders.

Figure 140. Radiocarbon dates from logs buried in glacial deposits in the upper Middle Fork valley above Ridley Creek. (Carrie Donnell)

Figure 141. Log buried in glacial deposits in the upper Middle Fork between the Deming glacier and Ridley Creek. (Photo by Carrie Donnell)

A lateral moraine high above the valley floor marks the ice margin of the Deming glacier during the Younger Dryas cold period (11,500 years ago) and the Little Ice Age about 500 years ago (Figs.142–144). Nested inside the highest lateral moraine are half a dozen younger moraines dating back to the Little Ice Age (400-500 years ago) (Fig. 145).

Figure 142. Lateral moraine high above the valley floor of the Deming glacier.

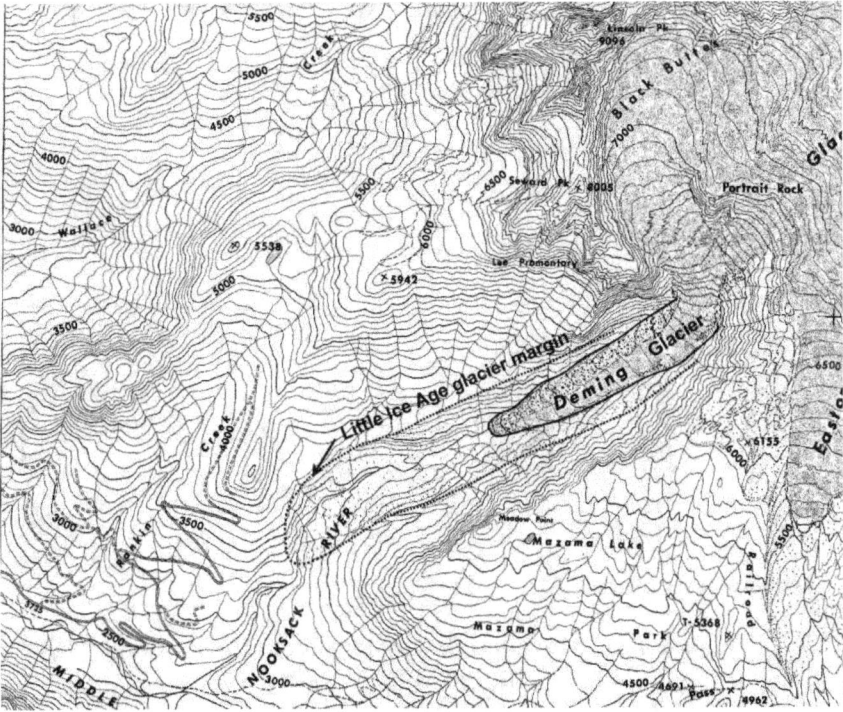

Figure 143. Former ice margin of the Deming glacier during the Little Ice Age. (Modified from USGS topographic map)

Figure 144. High lateral moraine and forest trimline in the lower Deming valley.

141

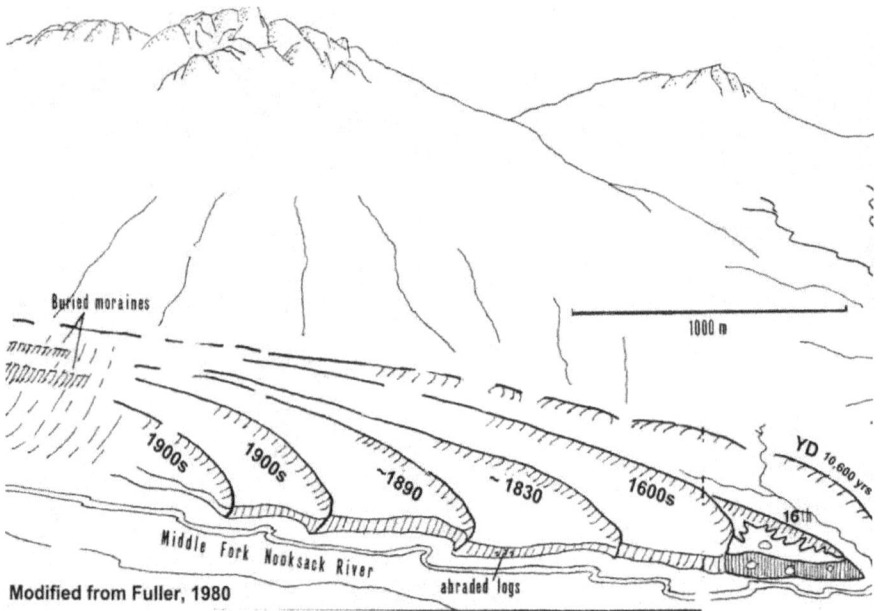

**Figure 145. Moraines downvalley from the Deming glacier.
(Modified from Fuller, 1980)**

**Figure 146. Debris accumulating at the terminus of the Deming glacier, building
an end moraine.**

Figure 147 shows the Deming glacier in 1915, well downvalley from the present terminus. Like the other Mt. Baker glaciers, the ice margin marks the position of the glacier terminus at the culmination of the global cool period from 1880 to 1915, considered to be the end of the Little Ice Age. The following warm period from 1915 to 1945 was the warmest period of the last century and the glacier receded more than a mile upvalley (Fig. 143). Following 30 years of cool climate and glacier advance (1945–1977), the global climate flipped abruptly from cool to warm in 1977, and the Deming glacier retreated about 1,000 feet upvalley by 2005.

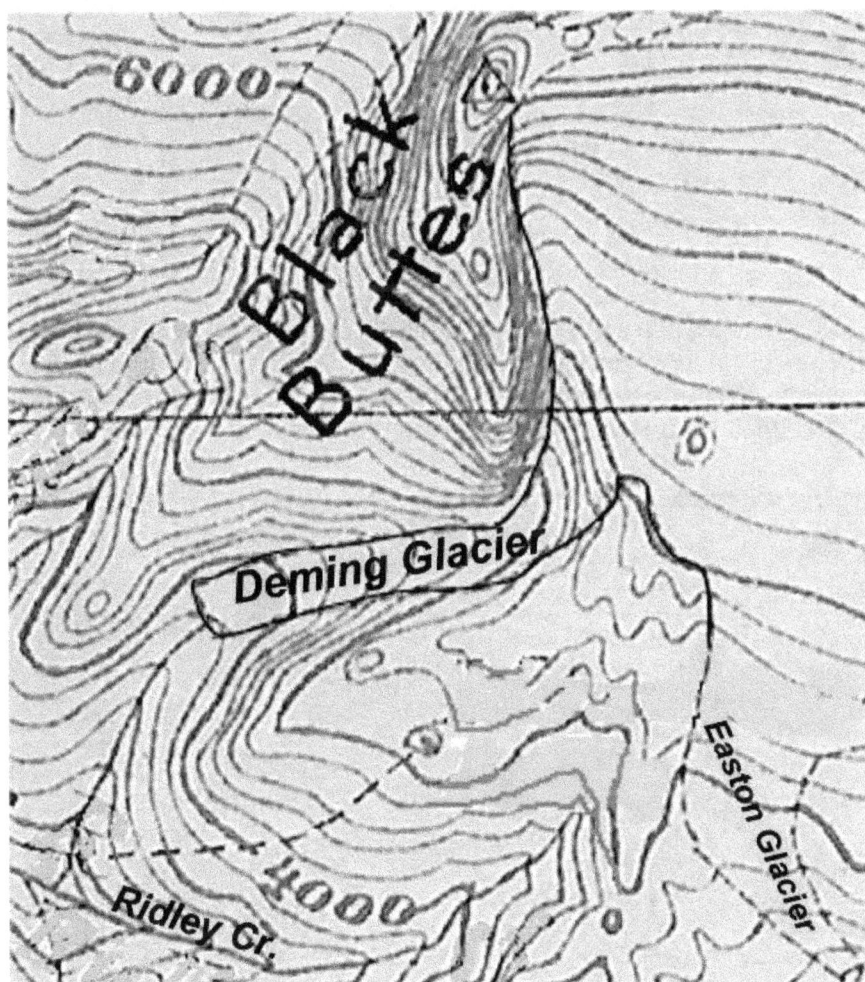

Figure 147. Deming glacier in 1915. (Modified from USGS topographic map)

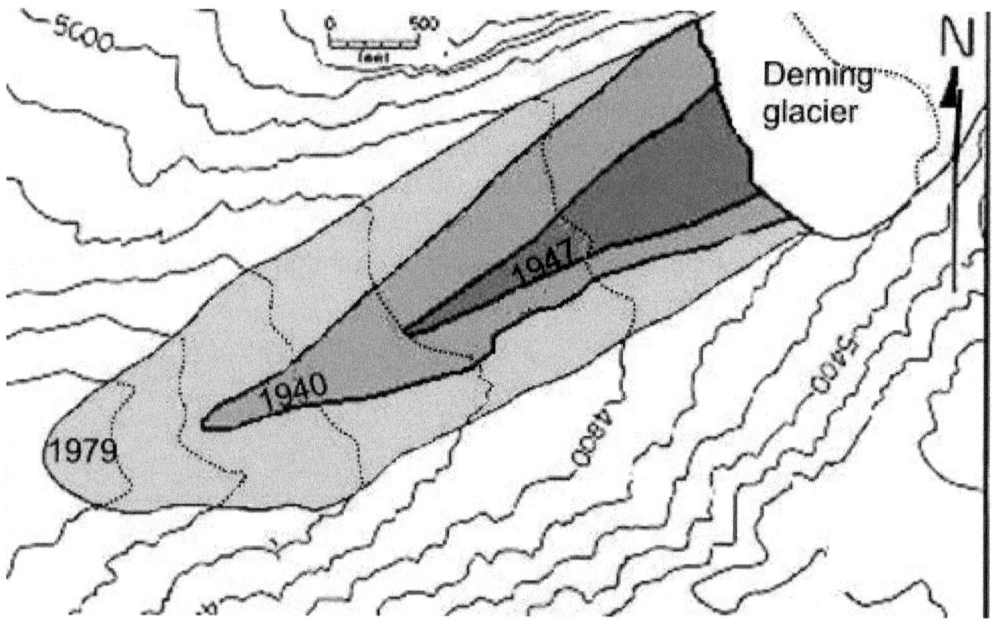

Figure 148. Changes in terminal positions of the Deming glacier 1940-47 near the end of the ~1915 to ~1945 global warming period. The 1979 terminus readvanced well downvalley from the 1940 terminus as a result of the ~1945 to 1977 global cooling. (Modified from Harper, 1992)

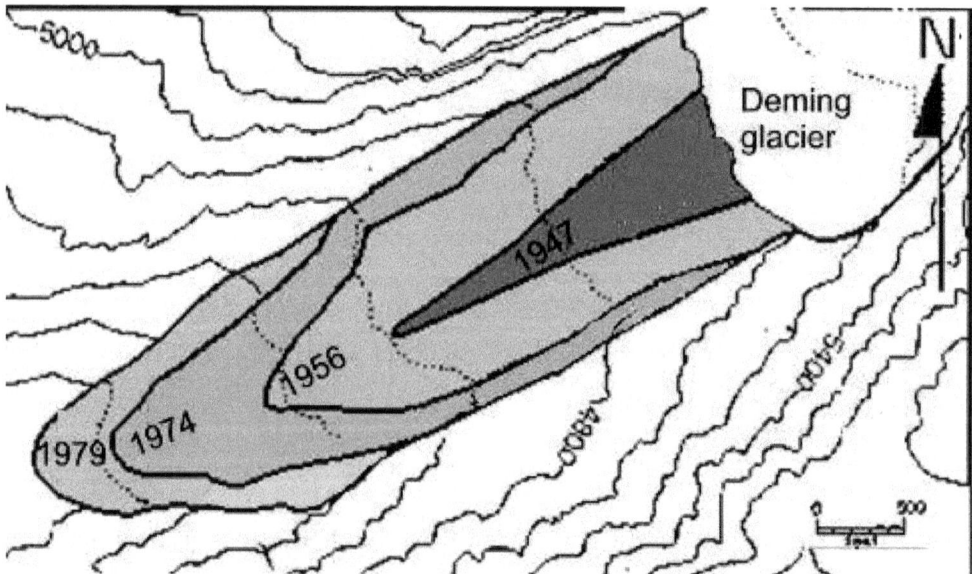

Figure 149. Advance of the glacier terminus downvalley from the 1947 terminus as a result of the ~1945 to 1977 global cooling. (Modified from Harper, 1992)

Easton Glacier

The Easton glacier originates high on the south flank of Mt. Baker below Sherman crater and extends downvalley to the headwaters of Sulfur Creek. During the Younger Dryas cold period of the last Ice Age, about 10,000 to 11,000 radiocarbon years ago the Easton glacier was much more extensive than now and covered most of the area between the present Easton and Squak glaciers. About 10,000 radiocarbon years ago, the global climate warmed abruptly, about 15° F in only 40 years, bringing the Ice Age to a sudden close. The climate remained warmer than present for the next 7,000 years and then entered a cooling phase about 3,000 years ago. During this cooling period, the global climate has oscillated between warm and cool at least 40 times in the last 500 years. During each cool phase, glaciers on Mt. Baker expanded and during each warm phase the glaciers retreated.

Figure 150. Easton and Squak glaciers on the south flank of Mt. Baker.
(Photo by Austin Post)

During the Younger Dryas cold period at the end of the last Ice Age (11-12,000 years ago), the Easton-Squak ice margin extended below timberline. Forested moraines now mark the former ice margins (Fig. 150). Radiocarbon dates from wood in a cirque moraine at Pocket Lake suggest that glaciers advanced again at 8,400 years ago.

145

Following a period of glacier recession during the Medieval Warm Period (900 AD to 1300 AD), global temperatures dropped abruptly during the Little Ice Age and glaciers on Mt. Baker expanded downvalley and buried forests growing on older moraines. The most prominent moraine, known as Railroad Grade, makes a ridge high above the valley floor (Figs. 150–152). It can be reached by hiking up the trail from Schreibers Meadow to Baker Pass and crossing the upper meadow to the trail along the crest of Railroad Grade. The drop down the inside of the ridge is precipitous and sod is often overhanging, so be sure to check before looking over the edge.

Figure 151. Little Ice Age moraines (Railroad Grade) along the Easton glacier valley. (Photo by Austin Post)

146

Figure 152. Railroad Grade, a large lateral moraine of the Little Ice Age Easton glacier.

Little Ice Age moraines east of Railroad Grade, known as the Medcalfe moraines, stretch across the broad, sloping bench to the Squak glacier (Figs. 150–155). They are composed of loose rocks and are barren of vegetation. Although no dates have been obtained directly from the moraines, they are correlative with other Little Ice Age moraines on Mt. Baker and are most likely about 500 years old.

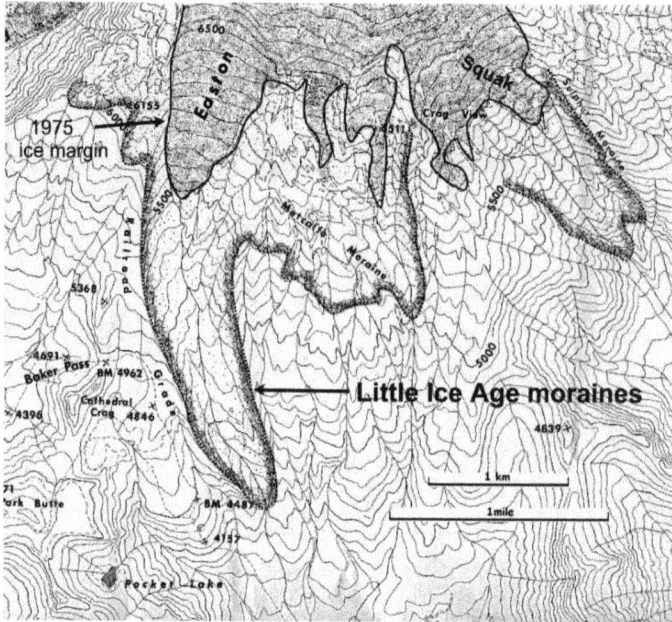

Figure 153. Little Ice Age (outermost) and 1975 glacier margins between the Easton and Squak glaciers. (Modified from USGS topographic map)

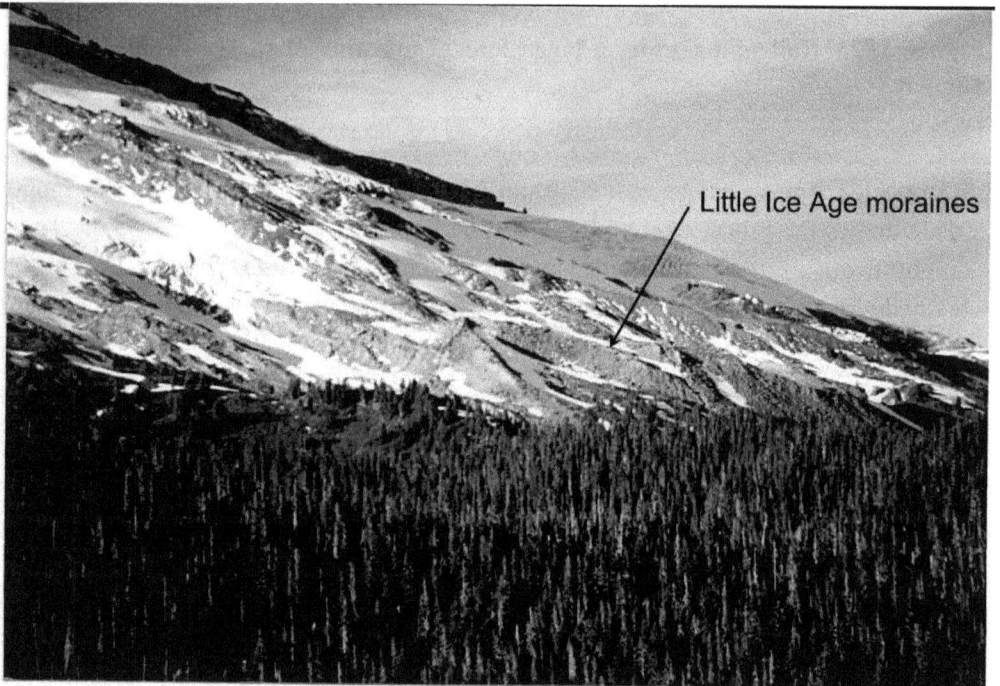

Figure 154. Little Ice Age moraines between the Easton and Squak glaciers.

148

Figure 155. Little Ice Age moraines between the Easton and Squak glaciers.

In 1915, the terminus of the Easton glacier was about a mile downvalley from the present terminus (Figs. 156–158) as a result of the global cool period from 1880 to 1915. This terminal position of the glacier was close to the maximum extent of the glacier during the Little Ice Age about 500 years ago. The global climate warmed from 1915 to 1945 and the Easton glacier retreated about a mile and a half upvalley.

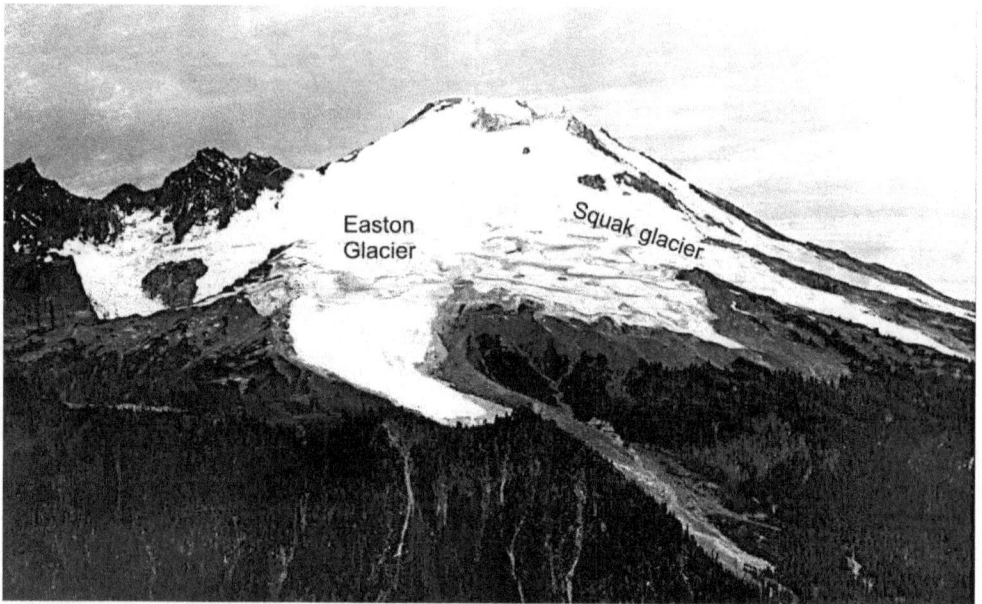

Figure 156. Easton and Squak glaciers, 1912, extending far downvalley. (Photo by Ed Walsh).

Figure 157. The Easton glacier early in the 1900s.

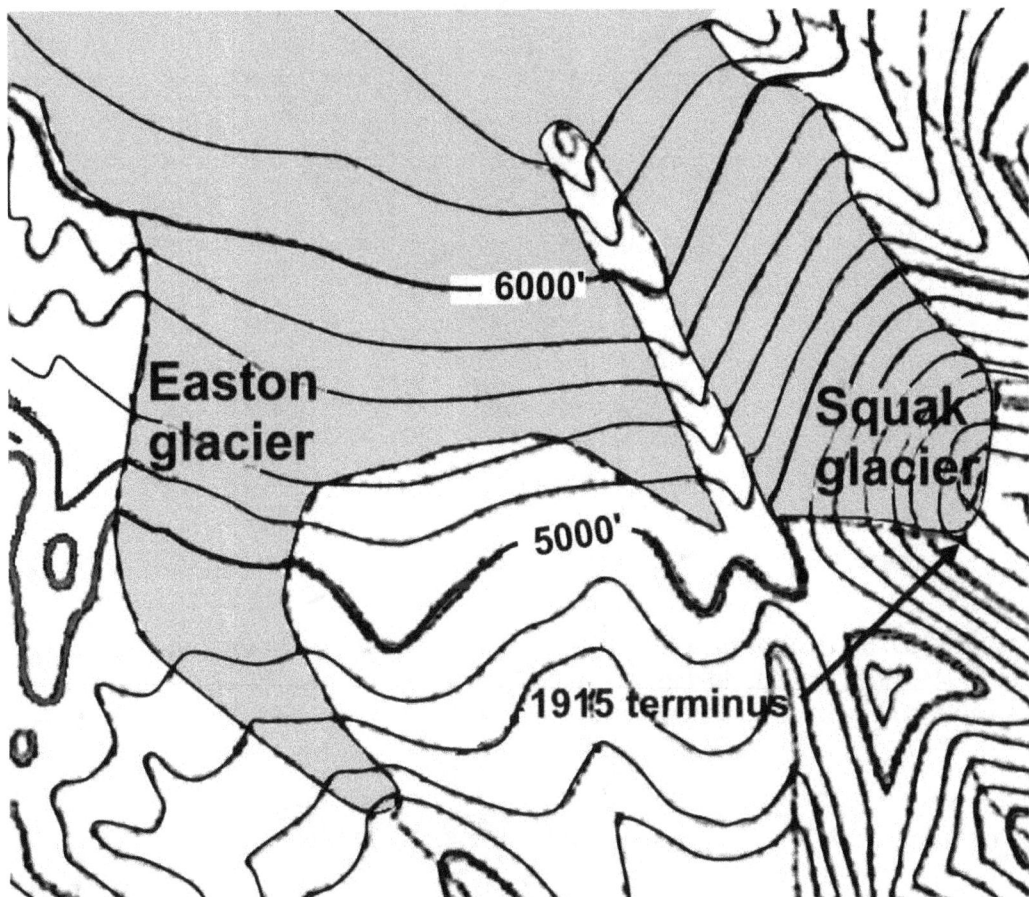

Figure 158. The Easton and Squak glaciers in 1915. The Easton glacier was about one mile below the present ice margin. (Modified from USGS topographic map)

The Easton glacier continued to retreat from its 1945 terminal position until the early 1950s (Fig. 159) then advanced from 1956 to 1979 (Fig. 160) when the Pacific Ocean and global climate cooled again and the glacier advanced about 1800 feet. In 1977, the Pacific flipped abruptly from its cool mode, where it had been since about 1945, into its warm mode, and the climate warmed again, inducing a glacier retreat of about 875 feet.

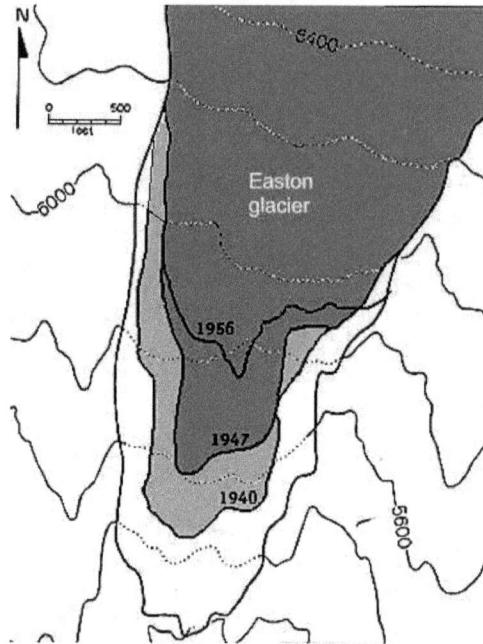

Figure 159. Retreat of the Easton glacier from 1940 to 1956 at the end of the ~1915 to ~1945 warm period, during which the glacier terminus retreated about a mile and a half upvalley from its 1915 position. . (Modified from Harper, 1992)

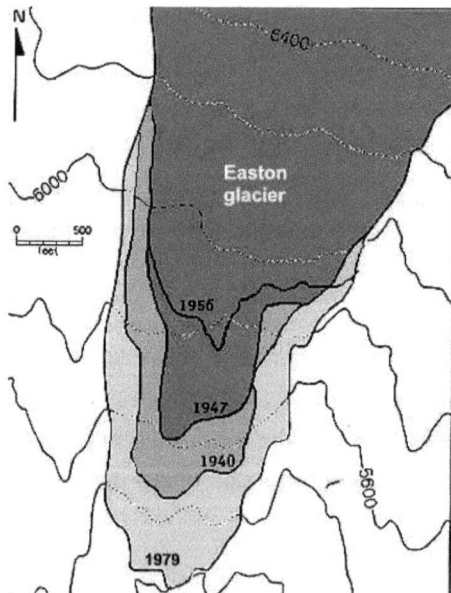

Figure 160. Advance of the Easton glacier from to 1956 to 1979 as a result of the ~1945 to 1977 global cooling. (Modified from Harper, 1992)

Boulder Glacier

The Boulder glacier originates at the eastern end of Sherman crater and extends downvalley to the headwaters of Boulder Creek (Fig. 161). During the Little Ice Age, about 500 years ago, the Boulder glacier extended about one mile farther downvalley from the present terminus (Fig. 162). A 1908 photo of the glacier (Fig. 163) shows thick ice well downvalley from the present terminus.

Figure 161. Boulder glacier. (Photo by Austin Post)

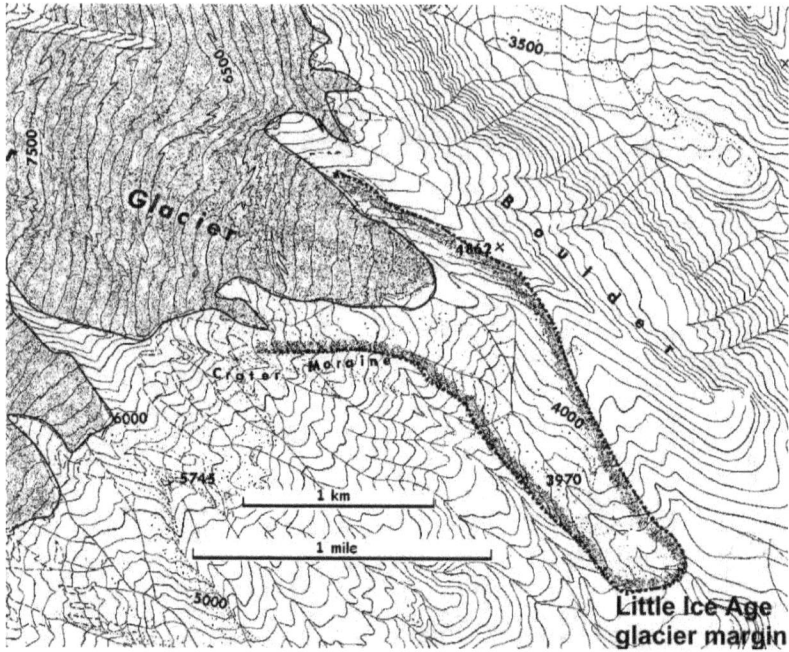

Figure 162. Little Ice Age and 1975 ice margins of the Boulder glacier. (Modified from USGS topographic map)

Figure 163. Thick ice of the Boulder glacier in 1908. The valley here is now ice-free. (Photo by Asahel Curtis)

154

**Figure 164. Forest trimlines left by retreat of the Boulder glacier.
(Tree ages from Burke, 1972)**

Several conspicuous trimlines in the forest near the position of the Little Ice Age glacier maximum can be used to date the retreat of the glacier (Fig. 164). Tree rings of the oldest trees unaffected by the glacier date back to 1558 AD. This date is probably close to the age of the Little Ice Age maximum of the Boulder glacier. Successive forest trimlines upvalley marking positions of former glacier termini date to about 1838, 1867, and 1920. In 1915, the glacier terminus was well downvalley from the present terminus (Fig. 165), but slightly upvalley from the Little Ice Age maximum position.

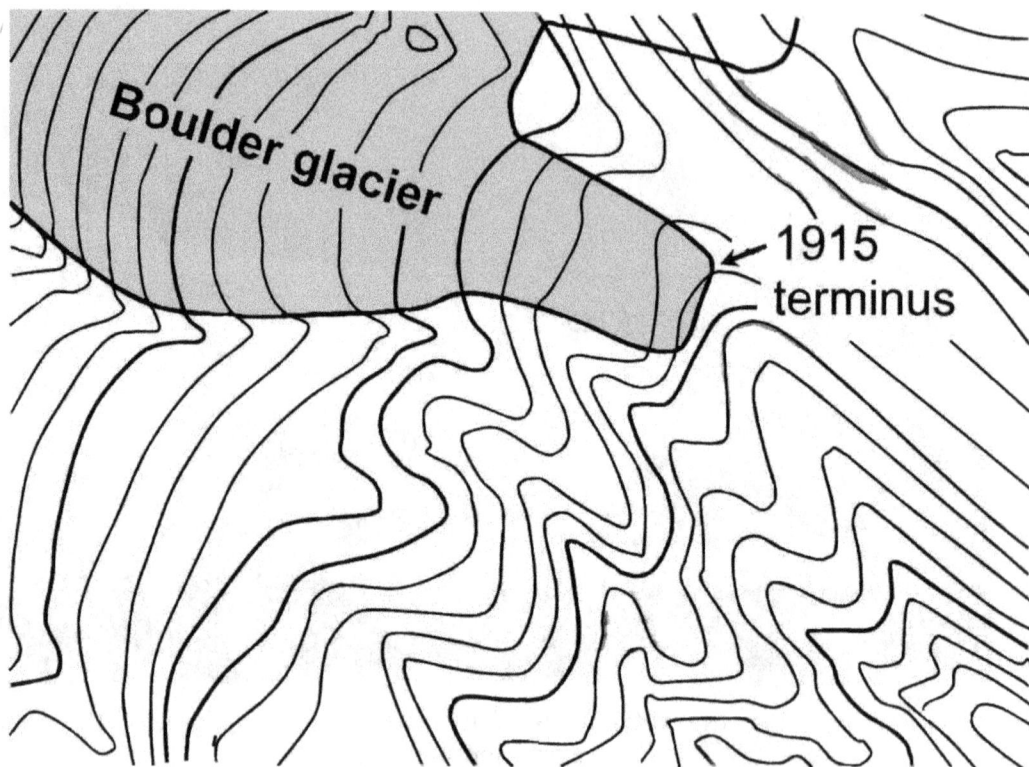

**Figure 165. Position of the 1915 terminus of the Boulder glacier.
(Modified from USGS topographic map)**

The 1915 terminus of the Boulder glacier (Fig. 165) marks the maximum extent of the Boulder glacier in this century as a result of the 1880 to 1915 cool period. From 1915 to 1945, the global climate warmed and the Boulder glacier retreated well upvalley. Figure 166 shows the position of the 1940, 1947, and 1956 termini.

The climate cooled from 1945 to 1977, but the Boulder glacier continued to retreat to the 1950s and then advanced vigorously about 2500 feet to its 1979 terminal position (Fig. 167). When the climate warmed again in 1977 (the Great Climate Shift), the Boulder glacier again was rather sluggish in its response and receded only about 80 feet by 1987. By 2005, the glacier had retreated about 1,500 feet.

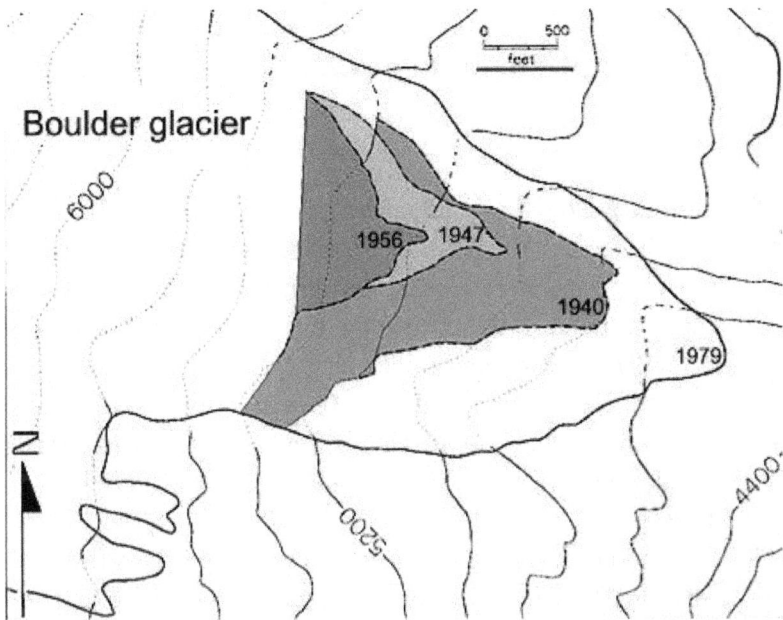

Figure 166. Terminal positions of the Boulder glacier during ice retreat from 1940 to 1956. (Modified from Harper, 1992)

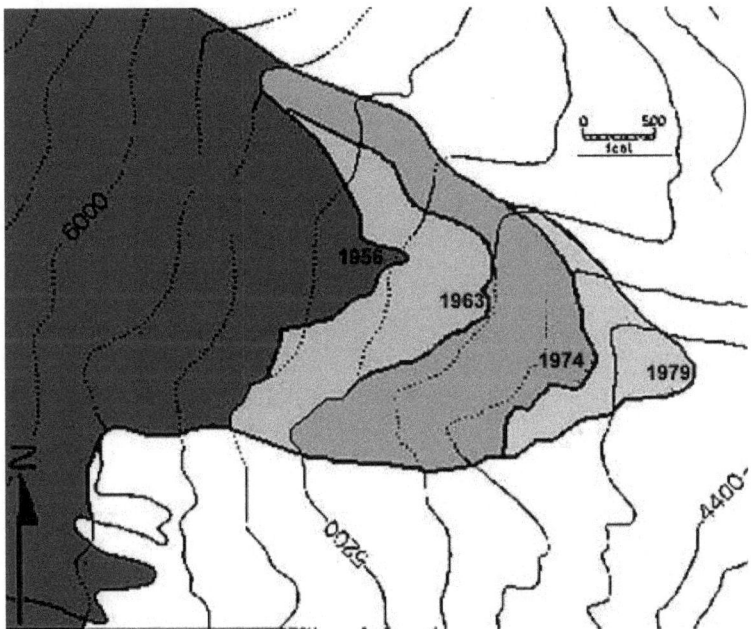

Figure 167. Terminal positions of the Boulder glacier during ice advance from

157

1956-1979. (Modified from Harper, 1992)

Implications of glacial fluctuations on Mt. Baker—what does it all mean?

Glaciers on Mt. Baker show a regular pattern of advance and retreat that matches sea surface temperatures in the eastern Pacific Ocean and global temperature changes. The Pacific Ocean has two temperature modes, a warm mode and a cool mode, and has historically switched back and forth between the two modes. In its cool mode, like the period from 1945 to 1977, ocean temperatures in the eastern Pacific were cool and Mt. Baker glaciers advanced. In its warm mode, like the period from 1915 to 1945 or 1977 to 1998, ocean temperatures in the eastern Pacific were warm and Mt. Bakers retreated (Fig.168). These mode switches are known as the Pacific Decadal Oscillation (PDO) and typically last 25 to 30 years.

The glacier fluctuations are clearly driven by changes in the PDO. An important aspect of this is that the PDO record extends to about 1900 but the glacial record goes back many years and can be used as an indicator of climate changes over many centuries. Extending this ongoing record into the future provides an opportunity to predict coming climate changes. We ended a 22 year warm period (1977 to 1998) when the PDO flipped into its cool mode in 1999. If he PDO behaves as it has through the past four previous global climate changes in the past century, we can look ahead to several decades of global cooling.

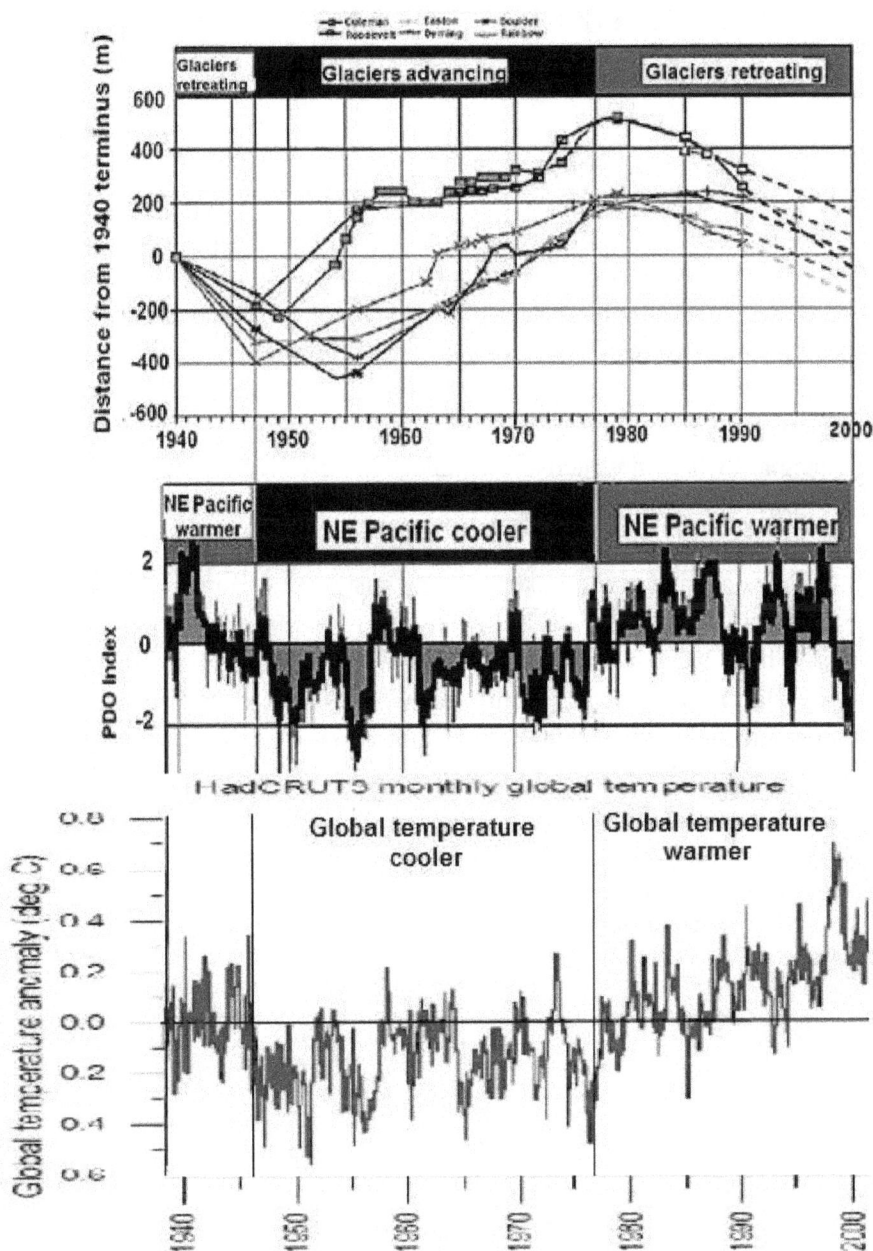

Figure 168. Relationship between Mt. Baker glacier fluctuations, the Pacific Decadal Oscillation (PD0), and global climate. When the PDO is cool, global temperatures are cool, and Mt. Baker glaciers advance. When the PDO is warm, global climate is warm, and Mt. Baker glaciers retreat.

VOLCANIC MUDFLOWS FROM MT. BAKER— A DANGEROUS VOLCANIC HAZARD

Volcanoes erupt large volumes of ash and fragmental material that accumulate on the flanks of the volcano. Volcanoes also expel large amounts of steam and hot water. As the steam and hot water migrate through the rock, they chemically alter and decompose it. Both the ash and chemically altered rock are unstable on steep mountain slopes and prone to slope failures, especially if eruptions of hot lava melt glacial ice, and produce volcanic mudflows (lahars) that can reach many miles downvalley where they can be very destructive. Lahars typically consist of water–saturated mud, ash and rock fragments, and trees swept up by the mudflows (Fig. 169).

An example of how destructive volcanic mudflows can be is the Osceola mudflow that came down from Mt. Rainier into the Puget Lowland about 5,000 years ago and left a deposit in the Seattle area about 100 feet thick. Similar, but somewhat smaller volcanic mudflows have occurred at Mt. Baker over the past several thousand years.

The Boulder Creek valley on the south flank of Mt. Baker is filled with volcanic mudflows, which have built a large delta into the Baker River valley, now largely covered by Baker Lake. Wood from one of the oldest of these mudflows has been radiocarbon dated at 8,700 ± 1000 radiocarbon years. Another old mudflow occurred at Schreibers Meadow where charred wood in a mudflow has been dated at 8,460 ± 140 and 8,500 ± 140 radiocarbon years.

160

**Figure 169. Volcanic mudflows from Mt. Baker at Schreibers
Meadow 8,500 and 5,800 years ago**

About 5,700 radiocarbon years ago (about 6,600 calendar years), a large mudflow extended at least 20 miles down the Middle Fork of the Nooksack River valley from Mount Baker. The mudflow, dated at 5,650 ± 110 radiocarbon years, was so thick it flowed up into Clearwater Creek, a tributary of the Middle Fork, reaching 130 feet above the level of the Middle Fork. Remnants of the mudflow also occur at the confluence of the Nooksack Middle and North Forks in the area near Welcome where it has been dated at 5,710 ± 110 radiocarbon years. . Many large logs in the mudflow are oriented parallel to one another pointing up the Middle Fork valley, indicating the direction of flow. Much of the wood here is charred, suggesting that the lahar may have been hot when it was

161

emplaced. Many of the rocks in the mudflow are rimmed with sulfur or are hydrothermally altered, indicating its volcanic origin. The Rocky Creek ash, dated at about 5,800 radiocarbon years, underlies the mudflow, suggesting that both are related to an eruptive phase of Mt. Baker. Most of the mudflow came down the Middle Fork valley, but it also spilled down the Sulphur Creek valley, on the south flank of the mountain where it forms an irregular surface on both sides of the Schreibers Meadow cinder cone and covers the Sulphur Creek lava flow and Schreibers Meadow scoria. Thin remnants of the mudflow occur on the divide between the Middle Fork and Schreibers Meadow, suggesting that the mudflows at both places belong to the same event.

Figure 170. Volcanic mudflow from Mt. Baker contain many logs at the junction of the North and Middle forks of the Nooksack River. Dating of the logs here gave an age of 5,710 ± 110 radiocarbon years.

162

Volcanic mudflows also came down the Park Creek valley about 6,000 radiocarbon years ago. A 15 foot thick, mudflow, containing hydrothermally altered rocks up to six feet in diameter is exposed on the north side of the road at lower Park Creek along the road to Baker Hot Springs, about half a mile from the junction with the Baker Lake Road. Wood from the mudflow has been dated at $6,650 \pm 350$ radiocarbon years. Another mudflow about 20 feet thick is exposed on the east side of the Park Creek fan near the mouth of Swift Creek. Abundant wood in it have been dated at $6,170 \pm 250$ radiocarbon years. A stump buried by a much younger mudflow along the banks of Park Creek has been dated at 530 ± 200 radiocarbon years.

A volcanic mudflow, exposed in the floor of the Nooksack Middle Fork channel about 15 miles downvalley from the headwaters, has been dated at $3,120 \pm 50$ radiocarbon years. This is the only known occurrence of the mudflow, but because of the distance downvalley from the source, it must have been a substantial flow.

Several volcanic mudflows in the past few hundred years occur in the Boulder Creek valley, and two large masses of rock debris have avalanched into the upper part of Rainbow Creek valley, leaving forest trimlines. The older of the two occurred in the mid-1860s and moved at least six miles downvalley. The younger event occurred in 1888, based on tree rings from an avalanche-damaged tree.

Volcanic mudflows from Mt. Baker pose the principal hazard to towns and farms along the Nooksack River. Lava flows are unlikely to extend far enough downvalley to impact populated areas, but subglacial eruptions that release large volumes of water can produce mudflows and floods that could be catastrophic.

CHAPTER 4

THE ICE AGE IN WHATCOM COUNTY—THE GREAT REFRIGERATION

The scenery of Whatcom County is notable both for its spectacular beauty and variety, from sandy and serene seascapes to tall timber and snow-covered mountains and glacier-clad volcanoes that rise two miles above sea level within only an hour's drive. In the North Cascades, many magnificent peaks provide a lifetime of challenges to avid mountaineers.

Uplift of the Cascades has sown the seeds of destruction of the range as the elevated mountains encourage erosive drainage off their high, steep slopes in an energetic attempt to make its ultimate rendezvous with the sea. Moreover, the high altitudes of the headwaters of the streams have been profoundly sculptured by modern alpine glaciers and the even larger glaciers of the Ice Ages. Today, the high peaks of the Cascades record the glacial origin in their scenery and continue to maintain active glaciers above 6,000 feet as reminders of the much more extensive refrigeration of the not-too distant past. During the Ice Ages, huge glaciers overwhelmed all but the very highest peaks in the county beneath thousands of feet of glacial ice. Today, great quantities of glacial debris from the Ice Ages, including much rock debris imported from Canada by the ice, record recurrent invasions and retreat of the glaciers.

During the last Ice Age (10,000 to 20,000 years ago), worldwide temperatures dropped dramatically and immense continental glaciers were built across vast areas of northern North America. Large accumulations of ice and snow in British Columbia resulted in the growth of an ice sheet, known as the Cordilleran Ice Sheet, that advanced southward into Washington at least six times (and probably more), only to melt away thousands of years later, leaving as their only footprint the glacial debris deposited by the ice. Sediments of each successive glaciation buried the previous ones so that only deposits of the last glaciation are exposed on the land surface of the Whatcom County, and evidence of the older glaciations may be seen only in deep sea cliff exposures.

Evidence of this invasion of northern ice may be seen today in the North Cascades where greenstones transported from Mt. Herman rest on the summit of Table Mt. (5,700 ft.) to the south, so the ice must have been slightly higher than 5,700 feet. Granite boulders are conspicuous as high as 5,300 feet on the rusty flanks of the Twin Sisters Range, which has no local source of granite bedrock, so the boulders must have been carried from British Columbia by glaciers whose surface was higher than 5,300 feet.

The huge size of these ice sheets dwarfs the small alpine glaciers which can be seen today in the Cascade Range and the ice which filled the Puget Lowland from the Olympics to the Cascades and from the Canadian boundary to Olympia was only a lobe of a still larger glacier complex in British Columbia. On a hot summer day, it is difficult indeed to imagine that ice more than a mile thick covered Bellingham 17,000 years ago (Fig. 171). Yet the deposits of rock fragments left by the ice give indisputable testimony that such immense glaciers occupied the lowland not only once, but at least six times and perhaps more. Periods of glaciation were separated by interglacial warm periods, climatically similar to or warmer than the present.

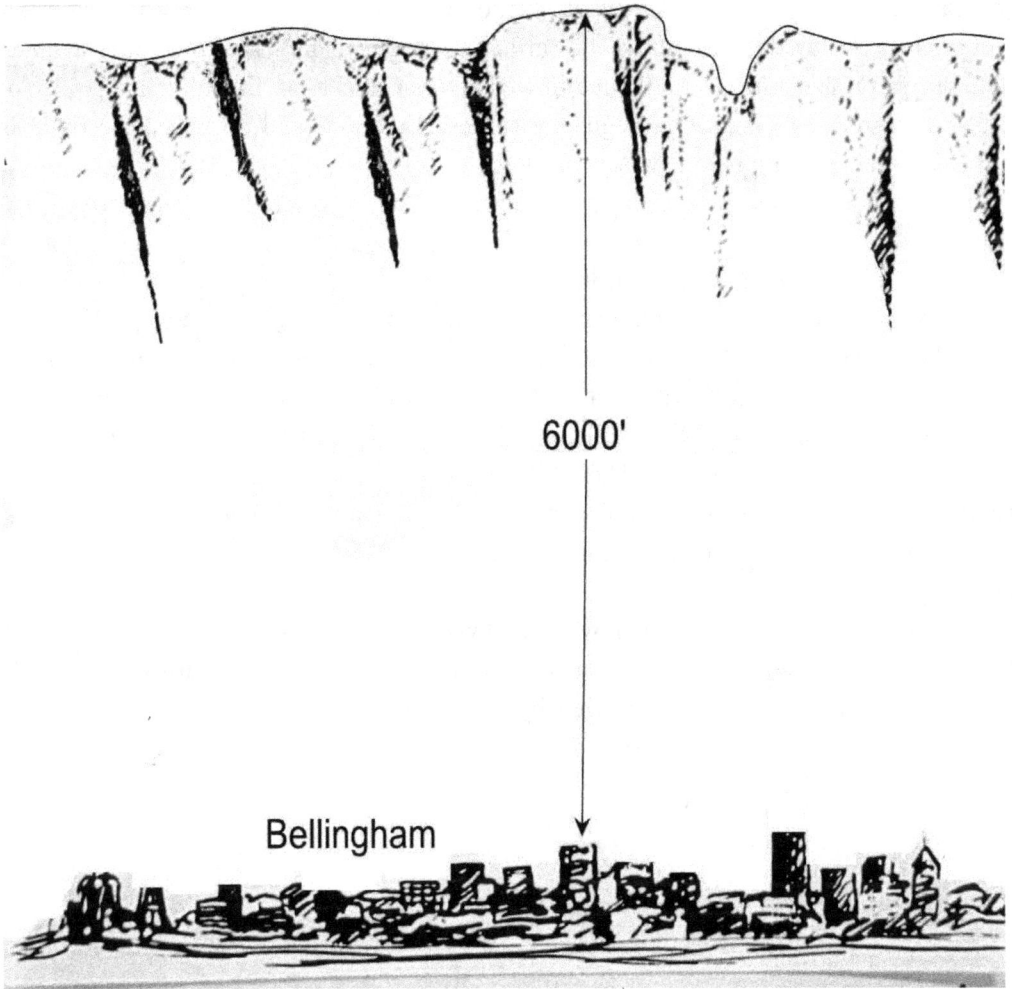

6000'

Bellingham

Figure 171. Bellingham was covered with over a mile of glacial ice 17,000 years ago.

Fraser Glaciation–Advance and retreat of the Cordilleran Ice Sheet

The last major phase of glacier growth in Whatcom County occurred during the Fraser Glaciation, which began about 20,000 years ago and ended 11,500 years ago. Evidence of three phases of the last glaciation, each known as a "stade," occur in Whatcom County.

During the oldest of the three stades, the Vashon Stade, growth of the ice sheet in the Puget Lowland reached its maximum extent south of Olympia, about 140 miles south of the Canadian border (Fig. 172). At that time, only mountain peaks in the Cascades 6,000 feet or more in altitude stood above the surface of the ice, as indicated by erratic boulders now strewn on their slopes. As the ice flowed across Chuckanut, Lookout, and Anderson Mts., it continued the scouring begun by earlier glaciations and deepened the basin of Lake Whatcom slightly below present sea level.

The Vashon glacier passed through Whatcom County sometime after 20,000 years ago and reached its maximum size between 14,000 and 15,000 years ago. As the ice thickened and moved southward, it inundated large areas of the northern Cascades near Mt. Baker and filled the lowland near Bellingham with ice more than a mile thick (Fig. 171). The direction of ice movement is shown by the numerous grooves and striations which occur on bedrock outcrops and by the elongate hills sculpted by flowing ice.

Rock debris carried by the Cordilleran Ice Sheet was deposited as a blanket of glacial till over much of the county. The till may now be recognized by its concrete-like appearance, with pebbles and cobbles imbedded in a matrix of clay, silt and sand (Fig. 173). It is well compacted as a result of the weight of several thousand feet of ice pressing on it.

As the Cordilleran Ice Sheet moved southward from Canada to the southern Puget Lowland, the rock debris it carried was ground up by crushing and shearing of the rocks by the ice. Many of the rocks carried by the ice were ground down to flat surfaces (faceted), polished, scoured, and scratched (Fig. 174).

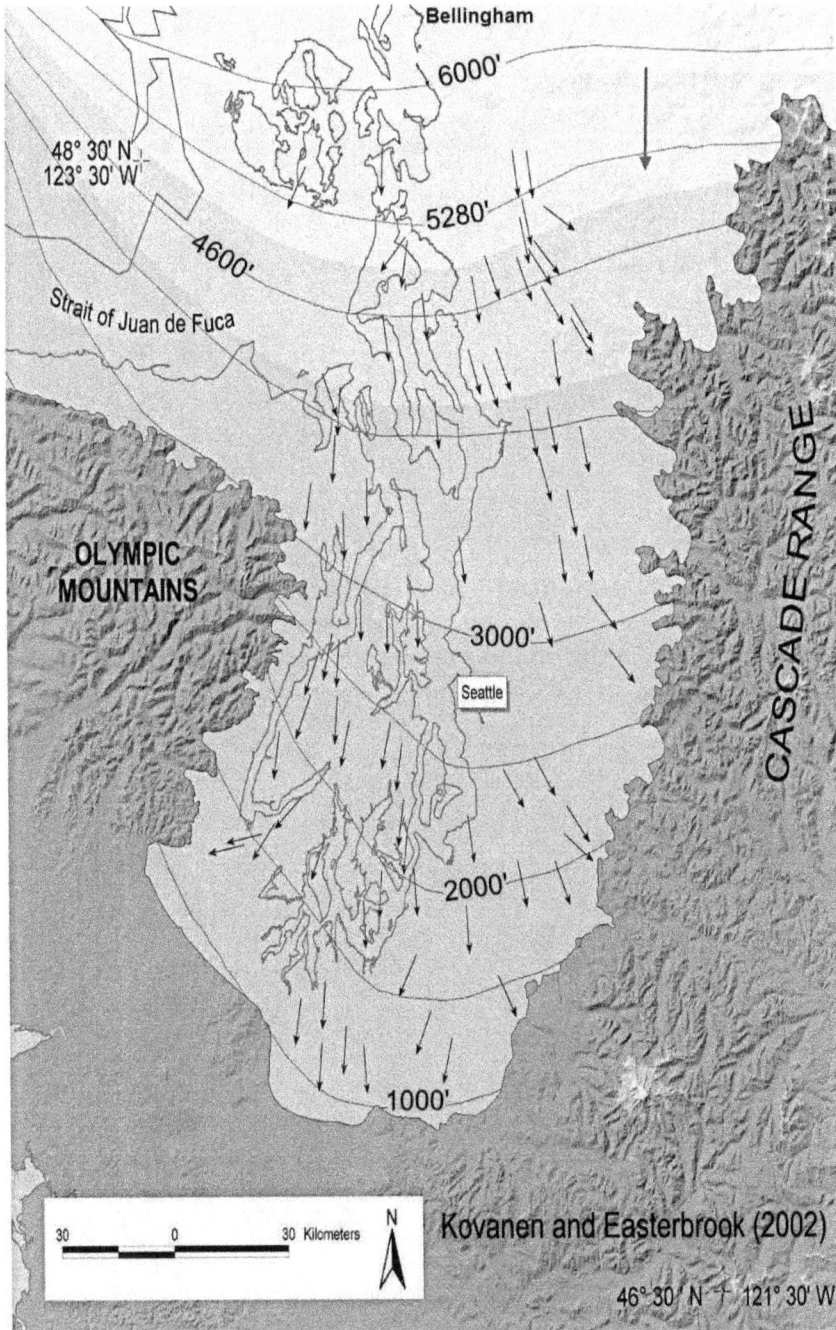

Figure 172. The Cordilleran Ice Sheet at the maximum of the last Ice Age. Numbers are altitudes of the ice surface above sea level. Lines are contours on the ice surface. Arrows indicate direction of ice flow. (From Kovanen and Easterbrook, 2002).

Figure 173. Vashon till deposited by the Cordilleran Ice Sheet about 15,000-20,000 years ago. The till has a concrete like appearance and is compact. The composition of pebbles and cobbles in the till indicate that they came from British Columbia.

Figure 174. Smooth, polished surface on a boulder carried by the Cordilleran Ice sheet and ground down to a flat surface. Note the many parallel scratches made by ice dragging rocks over the polished surface.

Since much of Vashon till at low elevations is covered with younger deposits, the best places to observe Vashon till are in sea cliff exposures such as north of Neptune Beach, north of Cherry Point, and south of Portage Point.

As the massive Cordilleran Ice Sheet moved over Whatcom County, it scoured the topography over which it was riding and produced streamlined hills and valleys parallel to the ice flow (Fig. 175).

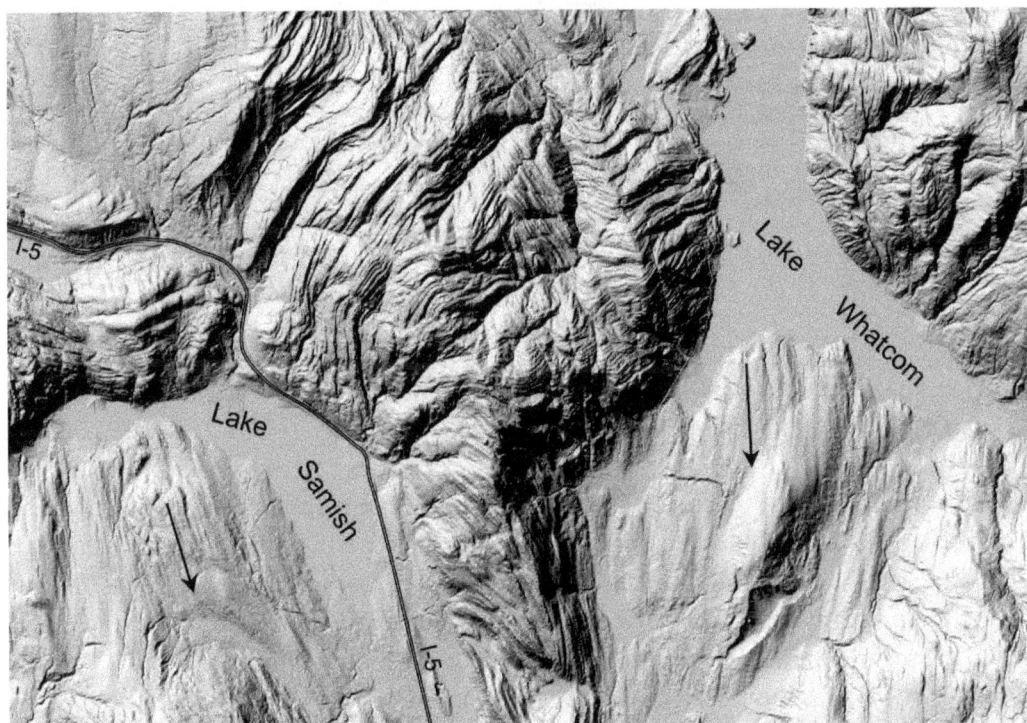

Figure 175. Lidar image of streamlining of topography overridden by the Cordilleran Ice Sheet. Arrows indicate the direction of ice flow. (USGS lidar image provided by Whatcom County)

As the Cordilleran Ice Sheet advanced southward from British Columbia, meltwater streams deposited sand and gravel in front of the ice margin. As the ice continued to advance and rode over the outwash sand and gravel, the force of the moving ice deformed the sediments (Fig. 176).

**Figure 176. Deformation of sand and gravel overridden
by the Cordilleran Ice Sheet.**

The second phase of the Fraser Glaciation is the Everson Interstade, during which the Cordilleran Ice Sheet thinned rapidly and disintegrated when marine water in Puget Sound floated the remaining ice (Fig. 177). The Puget Lowland as far south as Everett was covered with floating ice. As the ice melted, rock debris in the ice was released and fell to the seafloor below where it accumulated as glaciomarine stony clay.

The last stade of the Fraser Glaciation was the Sumas Stade when uplift of the land caused relative sea level to drop from about 500 feet to about 100 feet and the Cordilleran Ice sheet readvanced from the Fraser Valley near Sumas and covered most of the Whatcom County lowland with ice once again.

Rapid retreat and disintegration of the Cordilleran Ice Sheet

About 15,000 years ago, the massive Cordilleran Ice Sheet began to feel the effects of abrupt and intense climatic warming and underwent rapid, large–scale melting and recession. The ice sheet retreated northward by melting back of its terminus from its southernmost extent near Olympia and receded past the Seattle area. As the terminus of the glacier began to retreat, the ice also became much thinner. Relative sea level at that time was several hundred feet higher than at present (i.e., marine deposits of that age are now found well above sea level), and as the ice retreated to the vicinity of Whidbey

Island, it thinned enough that marine water entered the lowland from the Strait of Juan de Fuca and floated the ice, causing wholesale, rapid disintegration of the remaining ice sheet. Rapid disintegration of the Cordilleran Ice Sheet occurred in areas of deep water all the way to Canada and the region was covered with floating ice (Fig. 177). Ice remained locally grounded in some of the eastern part of the lowland and San Juan Islands. The disintegration of the Cordilleran Ice Sheet was probably similar to modern breakups of tidewater glaciers in Alaska. When glaciers reach a critical thickness, allowing marine to extend under the terminus, large masses of the lower glacier break off and float away from the terminus. This has happened many times to modern Alaskan glaciers, which serve as a modern analog to the breakup of the Cordilleran Ice Sheet.

Figure 177. Reconstruction of the disintegration of the Cordilleran Ice Sheet by floating of the ice in marine water about 13,000 years ago. The white areas represent disintegrating floating ice in sea levels higher than today. When the floating ice melts, rock debris is deposited on the sea floor as glaciomarine stony clay.

Everson glaciomarine interval—Submergence of the county beneath the sea ~12,000 years ago

After the Cordilleran Ice Sheet thinned, floated, and disintegrated, floating ice continued to deposit poorly sorted sediments on the sea floor, burying mollusk shells and other fossils in stony clay. Figure 178 shows the depositional environment of the sediments.

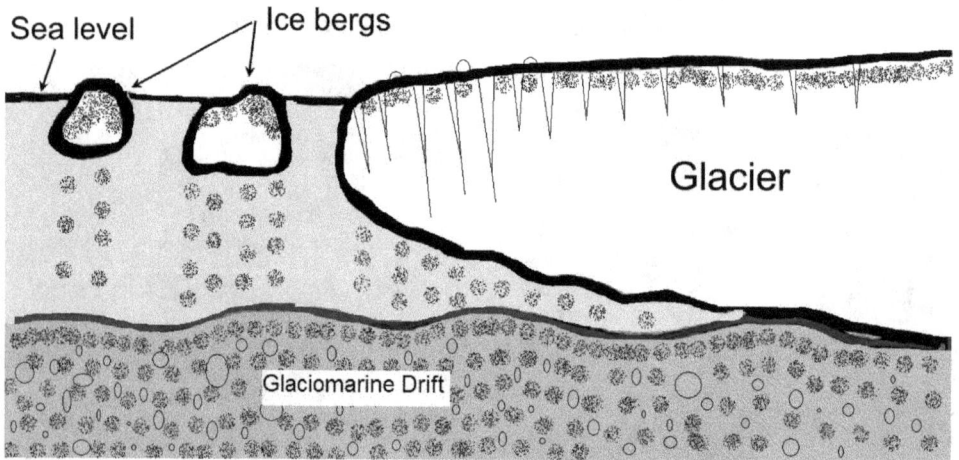

Figure 178. Depositional environment of glaciomarine sediments. As floating ice melts, clay, sand, pebbles, and cobbles in the ice are dumped on the sea floor, burying marine organisms on the bottom.

Debris released by melting of the ice sank to the sea floor below, depositing a heterogeneous mixture of clay, sand, pebbles, cobbles, and boulders resembling glacial till. Sedimentation rates were unusually high because of the large amount of rock debris released by melting of floating ice. Thicknesses of stony clay reached 70 feet and more in parts of Whatcom County.

Clams and various other mollusks living on the sea floor beneath the floating ice were buried in the deposits and remain today as fossils. Marine shells are widespread in these glaciomarine deposits, the most common being the clams Macoma, Chlamys, Saxicava, Nuculana and Saxidomus (Fig. 179).

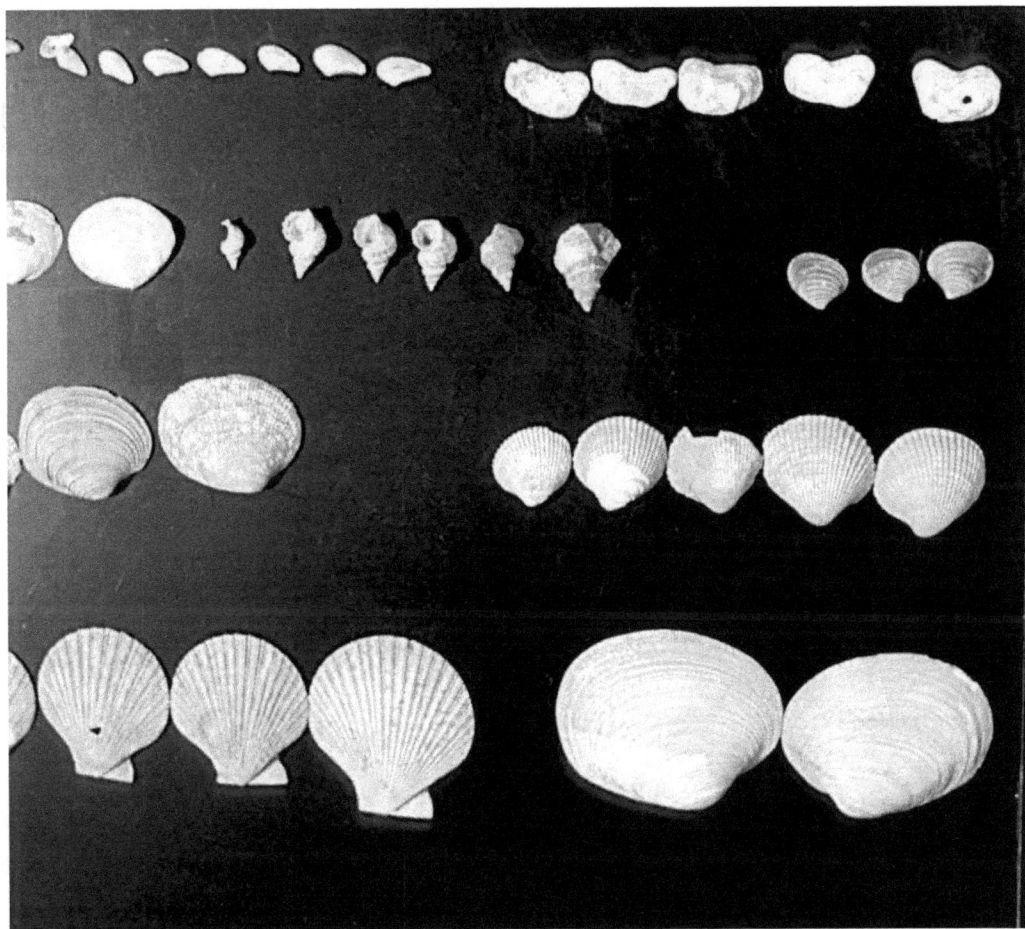

Figure 179. 12,000 year old fossil marine shells from glaciomarine stony clay in bluffs along Bellingham Bay.

At the highest sea level stand during the Everson glaciomarine interval, sea level was about 600 feet above present sea level and all of the lowland of Whatcom County was submerged beneath the sea (Figure 180).

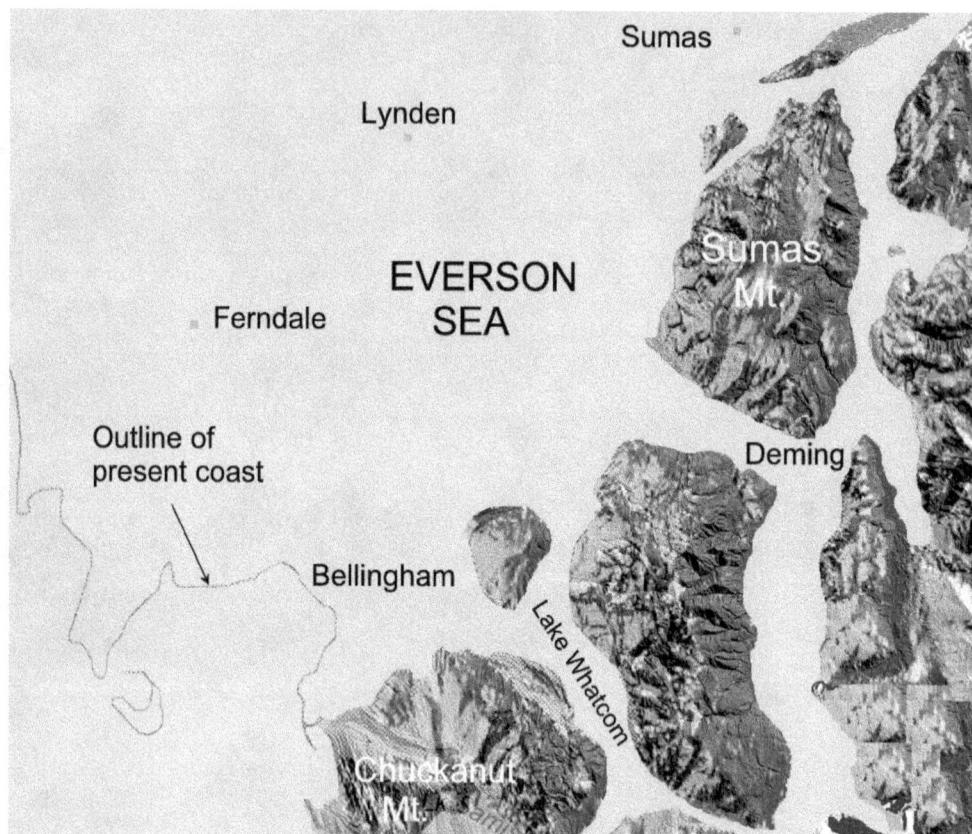

Figure 180. Reconstruction of what Whatcom County would have looked like during the Everson high sea level stand about 600 feet above present sea level 11,700 radiocarbon years ago. Most of the county was submerged and only the foothills stood above sea level.

The Everson glaciomarine enigma— the Whatcom County sea level yo–yo

Among the geologic mysteries of Whatcom County, none are more enigmatic than the relative sea level changes that occurred 11,700 to 12,500 radiocarbon years ago. In about 1,000 years, relative sea level in Whatcom County fluctuated from a few hundred feet above present at about 12,000 radiocarbon years ago to about 40 feet above present 11,800 radiocarbon years ago to ~600 feet above present sea level 11,700 radiocarbon years ago to near present sea level ~11,400 radiocarbon years ago. These rapid changes in sea level are some combination of rise and fall of sea level, uplift of the land, and crustal uplift.

174

However, any combination of these possible causes that explains the magnitude and abruptness of changes in sea level relative to land seems to be well beyond normal rates of crustal uplift. I discovered the evidence for these surprising changes in relative sea level back in the early 1960s, and over the past 50 years, no one has come up with either a tenable explanation for them or been able to refute the geologic evidence. The uplift rates necessary to explain the ups and downs of relative sea level seem implausible, but, as the saying goes, if it happened, it must be possible. The evidence for these remarkable relative sea level changes occurs in sediment sequences at several localities that mutually support one another and appears to be irrefutable.

Kulshan glaciomarine interval—the first piece of the puzzle

When the Cordilleran Ice Sheet melted dramatically in response to very abrupt global warming, the glacier thinned to the point where marine water entering the lowland from the Strait of Juan de Fuca caused the ice to float, resulting in sudden, wholesale disintegration of the glacier from Whidbey Island all the way into British Columbia. At that time, about 12,500 radiocarbon years ago, relative sea level was at least 300 feet higher than at present and a lot of debris in the glacier rained down on the sea floor as the floating ice melted.

Those deposits are preserved today as a geologic unit known as the Kulshan glaciomarine drift (stony clay), defined on the basis of sediments in the bluffs along the Nooksack River at the end of the Smith Road about 3 1/2 miles west of Deming (Fig. 181, 182).

The base of the Kulshan glaciomarine stony clay is exposed near river level where it is underlain by floodplain sand and silt containing two peat layers radiocarbon dated at about 12,000 radiocarbon years ago. These dates establish the age of the overlying Kulshan glaciomarine drift as younger than 12,200 years.

Figure 181. Sediments of the Everson glaciomarine interval at Nooksack River bluffs at the east end of the Smith Road, 3 1/2 miles west of Deming. Numbers are radiocarbon ages.

The 12,000 year–old floodplain sand is overlain by 40 feet of Kulshan glaciomarine stony clay deposited from floating ice when relative sea level was more than 300 feet above present sea level. Five radiocarbon dates from marine shells and wood in the Kulshan stony clay average 12,000 radiocarbon years.

The Kulshan glaciomarine stony clay is buried beneath younger sediments over most of the Whatcom County lowland. Exposures are known only from the Nooksack bluffs (Fig. 181), sea cliffs at Bellingham Bay (Fig. 183), road cuts at Deming (radiocarbon dated on marine shells at 11,970 radiocarbon years), and a stream gully at Welcome (dated from wood at 11,910 radiocarbon years).

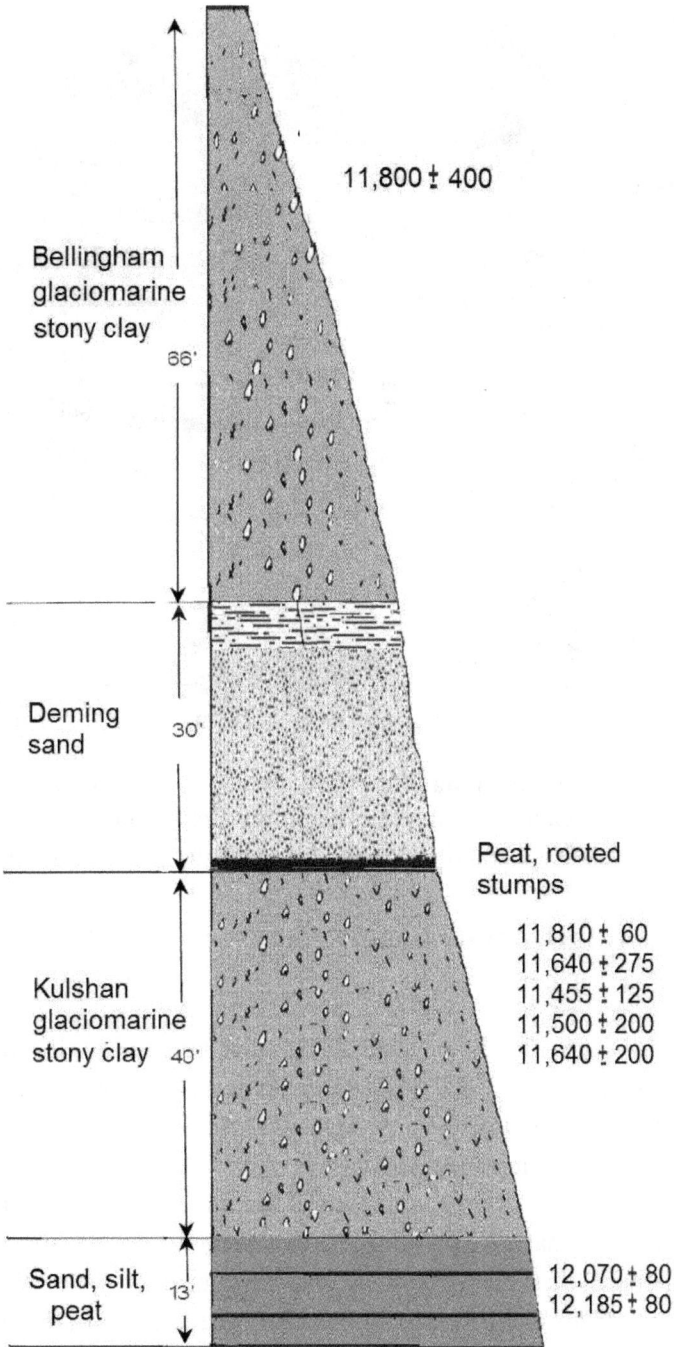

Figure 182. Sediments of the Everson interval in bluffs near the end of the Smith road. Numbers at the right are ages in radiocarbon years.

177

Deming beach sand

Kulshan glaciomarine deposits

Figure 183. Kulshan glaciomarine stony clay containing many marine fossils, overlain by Deming beach sand in sea cliffs along Bellingham Bay just below the old cement plant. The marine fossil shells shown in Figure 179 are from this locality.

Deming floodplain and beach sand—the second piece of the puzzle

Thirty feet of Deming floodplain sand overlies the Kulshan glaciomarine stony clay at the Nooksack River Bluffs (Fig. 181, 182). Cross–bedding and channels in the sand indicate that the Deming sediments were deposited on a floodplain.

A peat bed containing numerous rooted tree stumps occurs at the base of the Deming floodplain sand (Fig. 184). The significance of the rooted stumps is that they prove that following the Kulshan marine interval, the area emerged from the sea, allowing trees to grow. Radiocarbon dates on the stumps and other organic material indicate that the Kulshan interval ended about 11,800 radiocarbon years ago. Radiocarbon dates from wood and shells in Deming sand at Bellingham Bay are identical to those at the Nooksack locality.

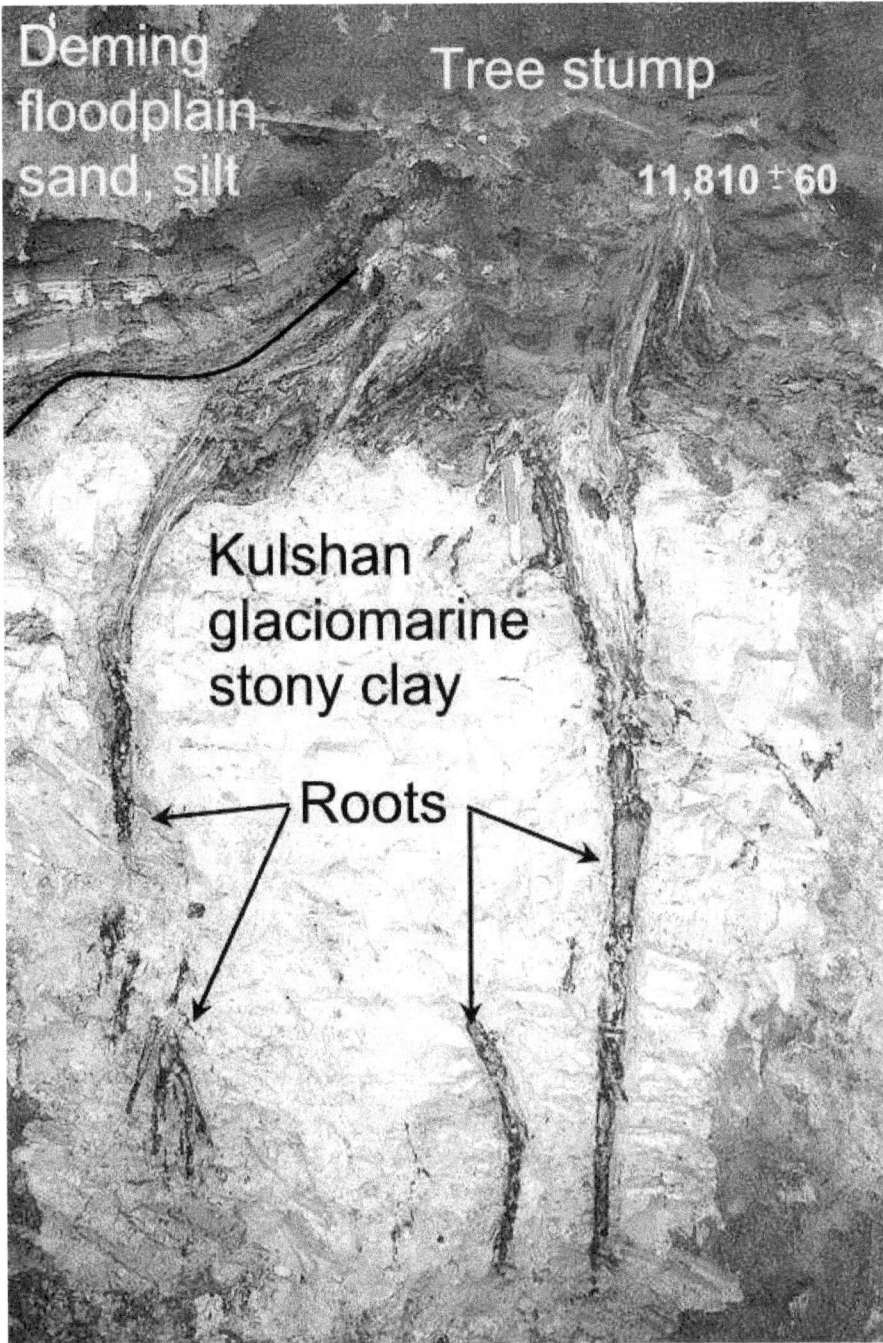

Figure 184. 11,810 year-old fossil tree stump in a peat bed at the base of the Deming sand. Nooksack River bluffs at the end of the Smith Road.

179

The Deming sand is also exposed in sea cliffs at Bellingham (Figs. 185, 186). Whereas the Deming sand at the Nooksack River bluffs bear a clear imprint of floodplain deposition, at Bellingham Bay, the Deming consists of beach sand and gravel containing abundant marine shell fragments. These beach deposits allow establishment of a sea level about 20–40 feet above present sea level at the time of deposition of the Deming sand. Considering that sea level during deposition of the underlying Kulshan glaciomarine stony clay was more than 300 feet above present sea level, that means sea level dropped close to 300 feet between deposition of the Kulshan and Deming units.

Figure 185. Deming beach sand overlain by Bellingham glaciomarine stony clay at sea cliffs along Bellingham Bay near Cliffside. This locality has abundant fossil marine shells in beach sand. Dating of shells and wood indicate an age of 11,800 radiocarbon years for deposition of the Deming sand.

BELLINGHAM
GLACIO-
MARINE
DRIFT

25'

Glaciomarine stony
clay, silt, and sand

Beach sand
and gravel   11,760± 85

Silt and clay

DEMING
SAND

34'

Beach sand with shells
and armored mud balls

Silt and clay

Beach sand with
many shells 11,685± 85

Poorly sorted silt, clay
and pebbles with shells

Covered by
vegetation

41'

KULSHAN
GLACIO-
MARINE
DRIFT

12,150±210
12,210± 80
12,900± 80

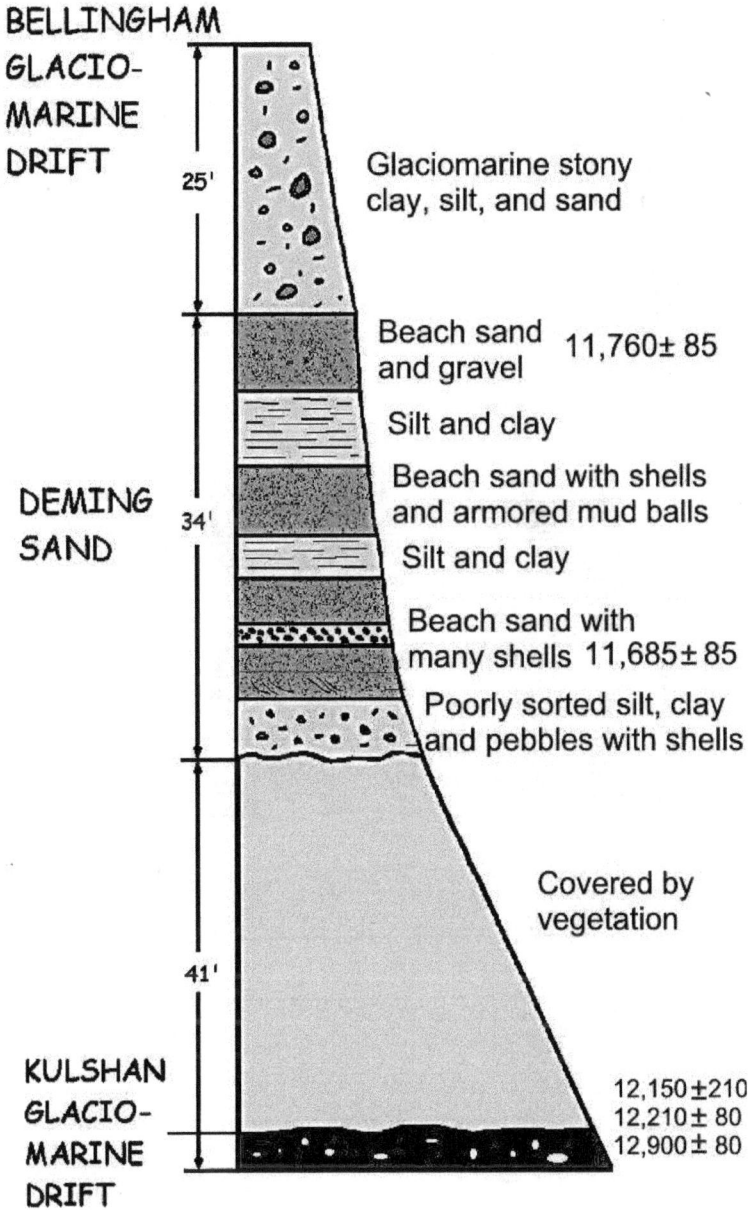

Figure 186. Sediments of the Everson interval exposed in sea cliffs at
Bellingham Bay near Cliffside. Numbers are radiocarbon ages.

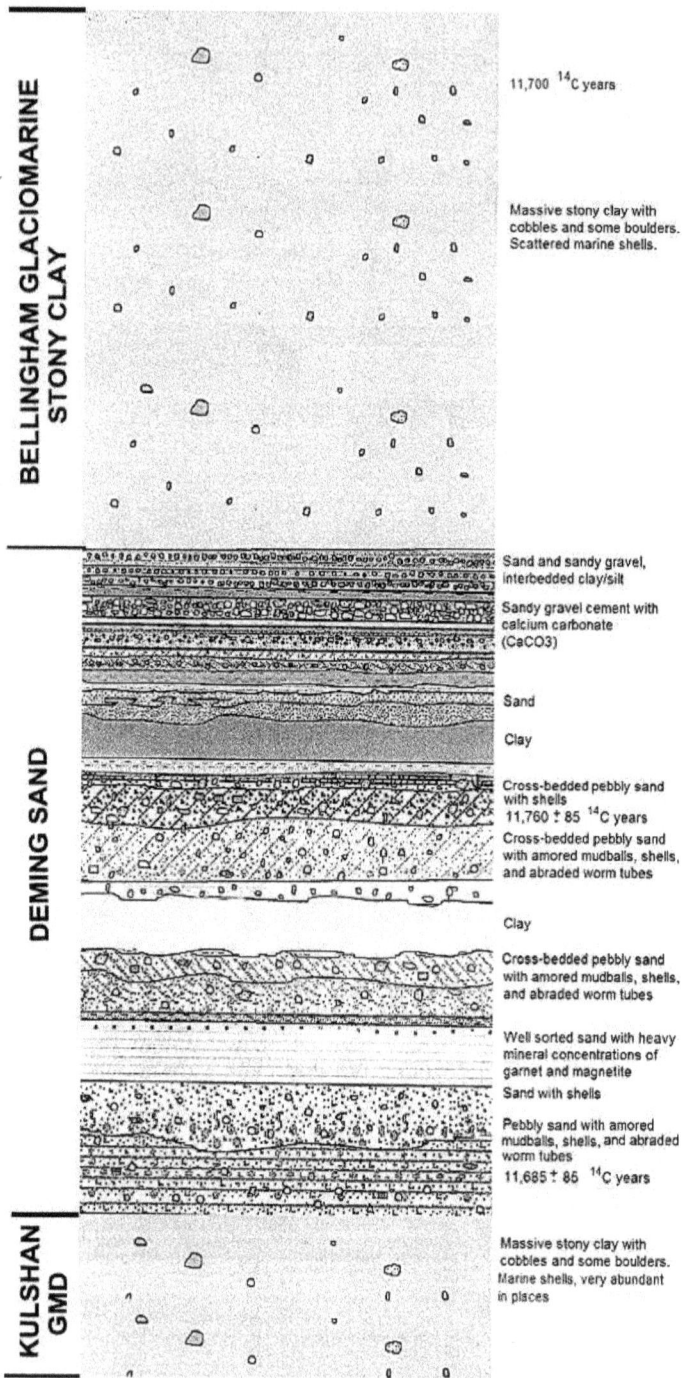

Figure 187. Sediments of the Everson Stade at Bellingham Bay sea cliff north of the old cement plant. (From Weber, 2001)

The Deming sand at Bellingham Bay sea cliffs shows many features typical of beach deposits—cross-bedding (Fig. 188), abundant wave-worn marine shells (Figs. 189, 190), heavy mineral concentrations of garnet and magnetite segregated by wave washing, shell 'hashes' made of ground up shell fragments (Fig, 189), and armored mudballs (Fig. 190). Armored mudballs form when blocks of clay slough off a bluff onto the beach and are rolled around by waves. Corners of the clay blocks are worn off and as the rounded clay balls roll around on beach sand, they coat the clay balls with an 'armor' of sand.

The significance of the Deming sand beach deposits at Bellingham Bay is that they indicate several hundred feet of emergence from the sea between deposition of the Kulshan and Bellingham glaciomarine sediments.

Figure 188. Cross-bedded, shelly, beach sand in Deming sand at a Bellingham Bay sea cliff north of the old cement plant (Photo by Stacy Weber).

183

Figure 189. Beach sand with abundant shells and ground up shell fragments belonging to the Deming sand unit in sea cliffs at Bellingham Bay near Cliffside. The ground up shells in the sand are produced by wave action and are identical to the modern beach sand at the base of the sea cliffs. Radiocarbon dating of the shells indicates an age of 11,760 years.

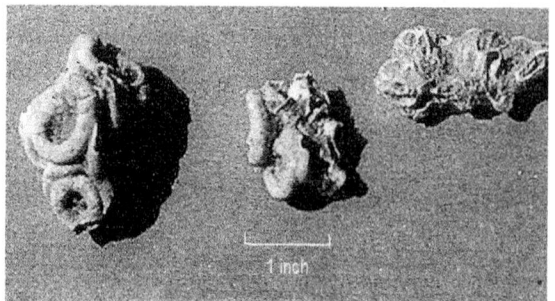

Figure 190. Armored mudballs (left) and wave–worn marine worm tubes (right) from beach deposits in the Deming sand in sea cliffs near Cliffside north of the old cement plant. (Photos by Stacy Weber)

184

Bellingham glaciomarine interval—the third piece of the puzzle

Following emergence of the Whatcom County lowland during deposition of the Deming sand, the lowland once again subsided below sea level. Once again the area was covered by the sea to elevations now 500–600 feet above present level and the water was crowded with floating ice bergs that dumped their load of debris on the sea floor when the ice melted. Sediment from the melting ice accumulated on the sea floor at an exceedingly rapid rate and the thickness of Bellingham stony clay at the Nooksack River bluffs (Fig. 182) reached at least 66 feet. About 25–35 feet of Bellingham glaciomarine stony clay was deposited over the entire lowland to elevations of at least 400 feet and probably as high as 600 feet above present sea level. We know that this sediment was glacial and that it came from British Columbia because many of the stones in the glaciomarine sediments have been glacially faceted (Fig. 191) and we know that they were deposited in marine water because marine fossils occur throughout the deposits (Fig. 192, 193).

The age of the Bellingham glaciomarine stony clay is well established at 11,700 radiocarbon years by 40 radiocarbon dates on wood and shells in the deposits. The astonishing thing about this age is that it is only about 100 years younger than the Deming sand, yet in that short period of time, the lowland was submerged 500–600 feet and 35-65 feet of glaciomarine stony clay was deposited over all of the lowland.

Figure 191. Glacially faceted and polished stones from Bellingham glaciomarine stony clay.

Figure 192. Articulated clam shells in glaciomarine stony clay. The clams were living on the sea floor when buried by rock debris released from the melting of floating ice.

Figure 193. 11,700–year old barnacle fossil attached to a glacially faceted pebble in Bellingham glaciomarine stony clay.

Figure 194 shows a cross–section of the sediments of the Everson glaciomarine interval. The same Kulshan–Deming–Bellingham sequence exposed at the Nooksack River bluffs is replicated in sea cliff exposures at Bellingham Bay (Figure 186), in a deep road cut at Deming, and in at least four localities in British Columbia. Radiocarbon dating of wood and shells at all of these localities establishes (1) the regional extent of the two glaciomarine units separated by Deming floodplain sediments, (2) the age of the sediments, and (3) relative sea level changes in the area.

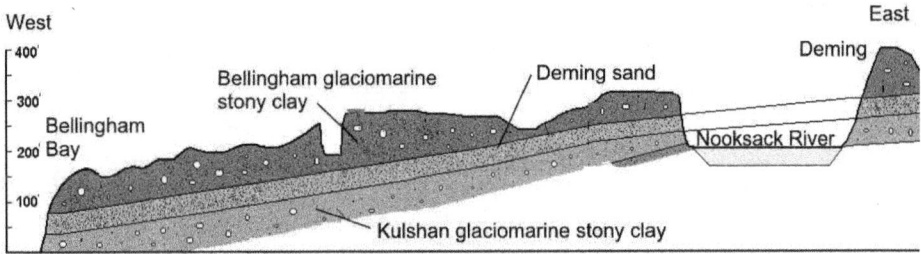

Figure 194. Geologic cross–section from Bellingham Bay to Deming showing the sediments of the Everson glaciomarine interval.

Old shorelines—footprints of ancient sea levels

The Bellingham glaciomarine interval is well dated at 11,700 radiocarbon years ago. Marine shorelines younger than the Bellingham glaciomarine interval occur at 540 feet above present sea level east of Blaine (Fig. 195), and many successively lower shorelines (Figs. 195, 196) record the emergence of the lowland from the sea. Similar multiple fossil shorelines occur on the western part of the Lake Terrell upland between Ferndale and Georgia Strait, on the Lummi Peninsula, on Lummi Island, and on Portage Island. A date of 11,400 radiocarbon years from the base of a peat bog in an abandoned channel just west of the Aldergrove customs station at the Canadian border indicates that the lowland had re–emerged by that time and sea level was less than 100 feet above present sea level by 11,400 years ago.

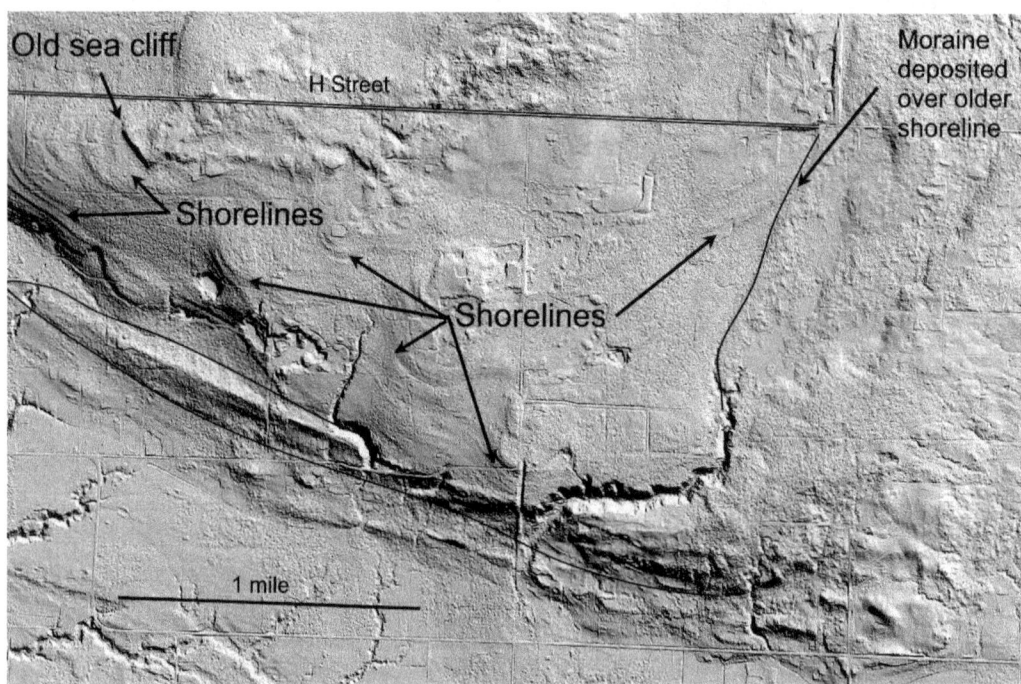

Figure 195. Marine shorelines and old sea cliffs up to 500 feet above present sea level, just east of Blaine. The old shorelines are covered by a moraine on the right side of the photo, showing that the moraine is younger. (USGS lidar image provided by Whatcom County)

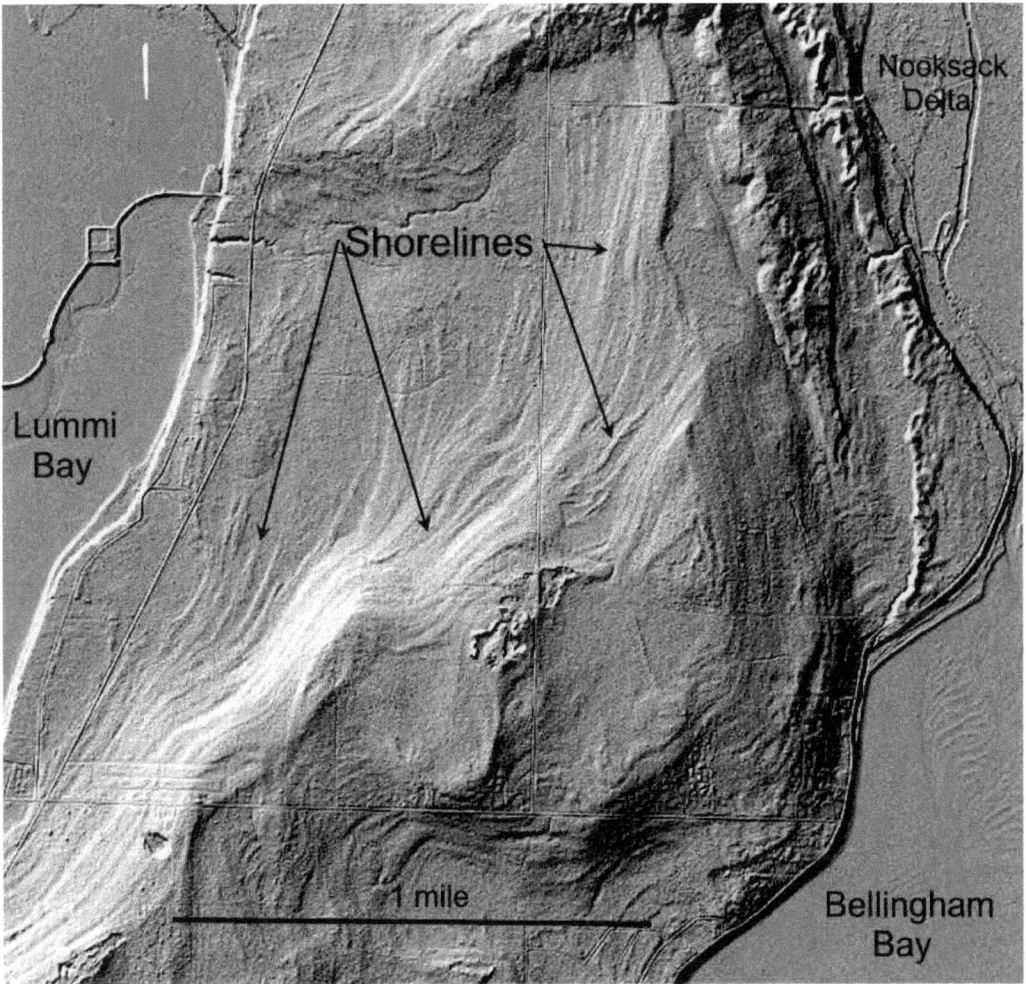

Figure 196. 11,400 to 11,700 year old fossil shorelines on the north end of Lummi Peninsula. (USGS lidar image provided by Whatcom County)

These uplifted shorelines record the emergence of the lowland from beneath that sea at the end of the Everson glaciomarine interval. The highest shoreline is at least 540 feet above present sea level and is probably the highest sea level reached during deposition of the glaciomarine sediments. The emergence of the lowland completes the last phase of the Whatcom County sea level yo-yo. The evidence for the abrupt ups and downs of relative sea level is conclusive, but the cause is elusive because the rates of change are well beyond rates known from elsewhere. The yo-yo doesn't occur south of Whatcom County and is apparently a local phenomenon.

Fraser Glaciation	Sumas Stade	**Sumas stade: (~11,400 - 10,000 ^{14}C yrs** • Ice sheet margin fluctuates • Relative sea level drops emergence of land
	Everson Interstade	**Bellingham glaciomarine drift: (11,700 ^{14}C yrs** • Marine waters reached to 600 ft. • Second submergence of land • Relative sea level rises again
		Deming sand: (~11,800 ^{14}C yrs.) • Relative sea level 30 – 70 ft. • First emergence of land • Relative sea level drops due to rising land surface • Land rises (isostatic rebound)
		Kulshan glaciomarine drift: (12,210 ^{14}C yrs • Relative sea level rises ~ 380 ft. • First submergence of land • Depressed lowland • Melting and thinning of ice sheet
		Vashon Stade: (~18,000 - 13,000 ^{14}C yrs) • Retreating and thinning of ice sheet allowed entrance of marine waters into Strait of Juan de Fuca • Ice sheet ~6000 ft thick in Whatcom County • Relative sea level lower than present day sea level

Figure 197. Events of the Fraser Glaciation.

READVANCE OF THE CORDILLERAN ICE SHEET 10,000–11,500 YEARS AGO—THE SUMAS MORAINES

Very soon after the lowland emerged from the Everson glaciomarine interval, the Cordilleran Ice Sheet re–advanced across the county from its source in the Fraser valley of British Columbia northeast of Sumas. The earliest re–advance reached almost to Blaine where it buried older raised shorelines (Fig. 195) made during emergence from the 540 foot sea level stand. As the Cordilleran Ice Sheet moved over the land surface, it rode over the previously deposited Bellingham glaciomarine stony clay and the moving ice shaped the surface into streamlined forms (Fig. 198). This re–advance is known as the Sumas Stade (a stade is a time of glacier advance).

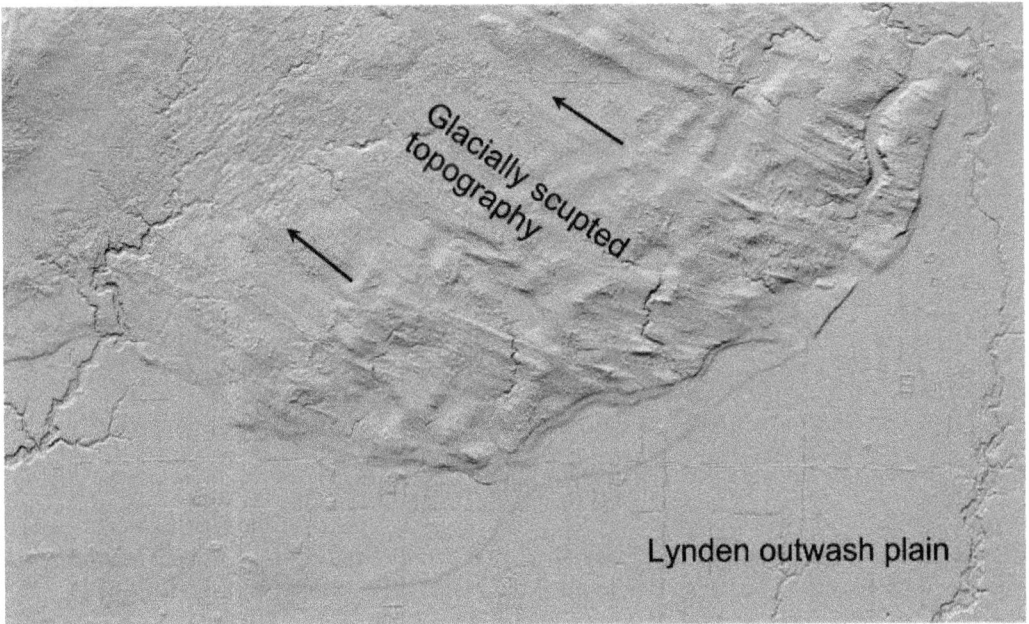

Figure 198. Streamlined topography made by scouring of the Cordilleran Ice Sheet during the Sumas advance between 11,400 and 11,700 years ago. The glacier moved from right to the upper left. (USGS lidar image provided by Whatcom County)

191

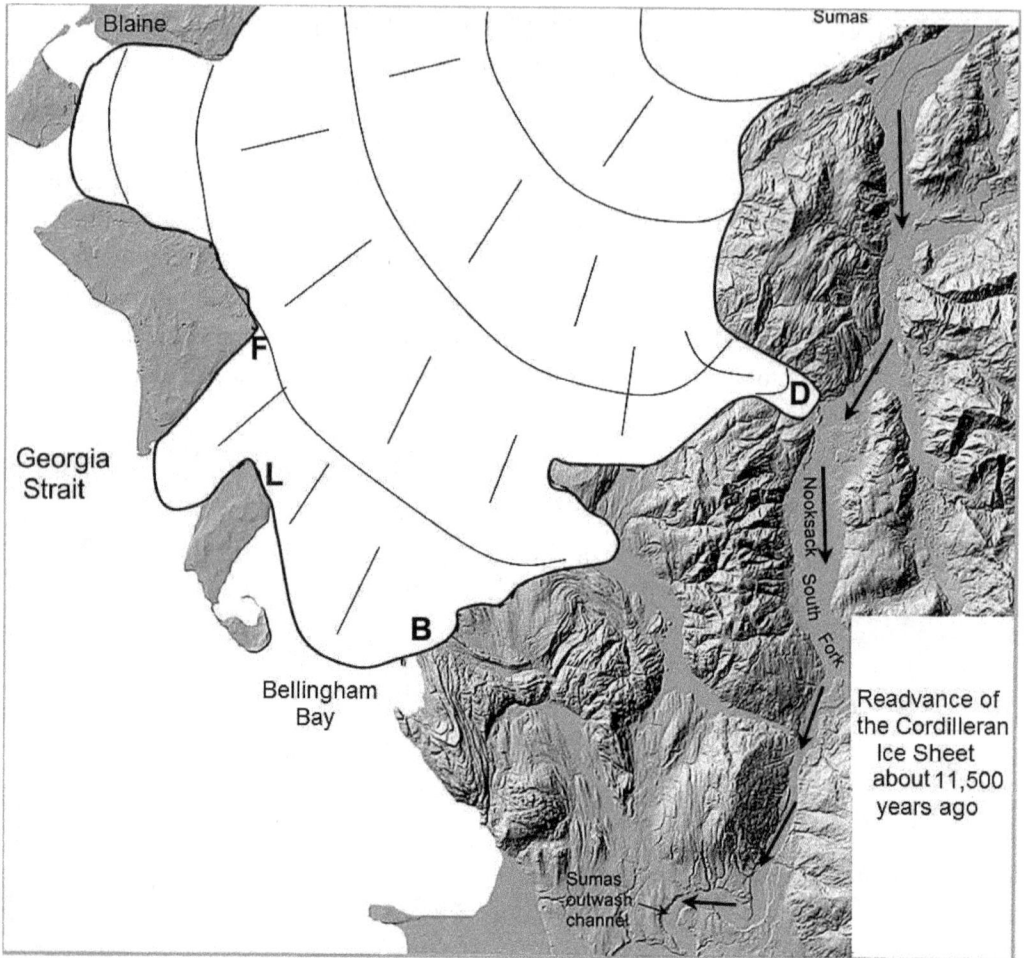

Figure 199. Reconstruction of the early Sumas readvance of the Cordilleran Ice Sheet about 11,500 years ago. F=Ferndale, L=Lummi Peninsula, B=Bellingham, D=Deming. (Base image USGS lidar provided by Whatcom County)

Figure 199 shows a reconstruction of the extent of the Cordilleran Ice Sheet remnant during building of the outermost Sumas moraines. The glacier came out of the Fraser Valley at Sumas and spread over the lowland as a piedmont lobe that deposited prominent moraines east of Blaine (Fig. 195), west of Ferndale (Fig. 201), the Lummi Peninsula (Fig. 196), and blocked the Nooksack River valley near Deming, forcing the Nooksack drainage southward through the South Fork valley and the Samish River where it built a delta into the Skagit Valley (Fig. 200).

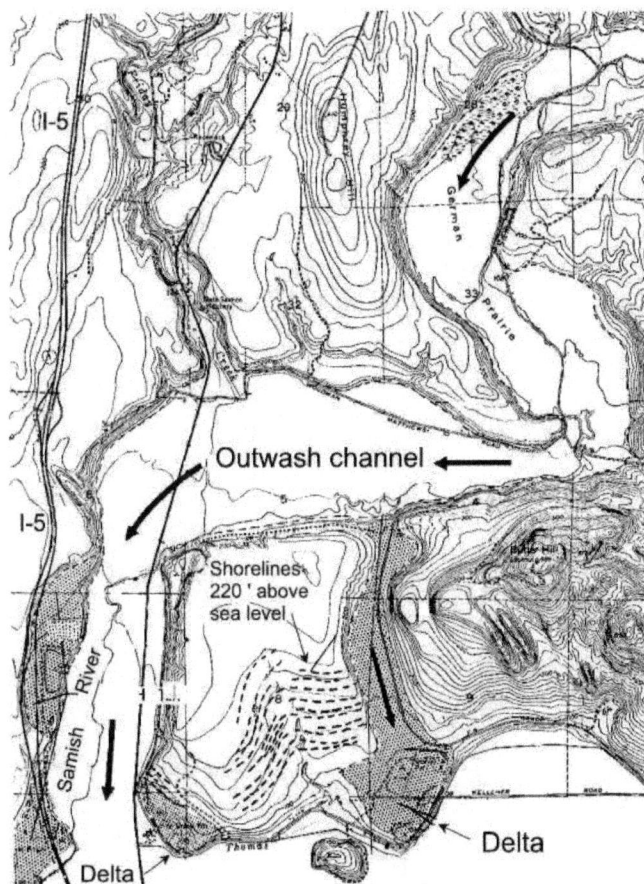

Figure 200. Samish River outwash channel. When the Cordilleran Ice Sheet of the Sumas Stade blocked the Nooksack River valley at Deming about 11,400 radiocarbon years ago, meltwater flowed through the Nooksack South Fork valley into the upper Samish River valley and built a delta into the Skagit Valley about 100 feet above present sea level. The bends in the valley are much wider than those of the present Samish River so the meltwater discharge must have been considerably higher than the modern Samish River. The meltwater channel is cut into Everson glaciomarine stony clay and the channel truncates multiple shorelines between 100 and 250 feet above present sea level. (From Easterbrook, 1994)

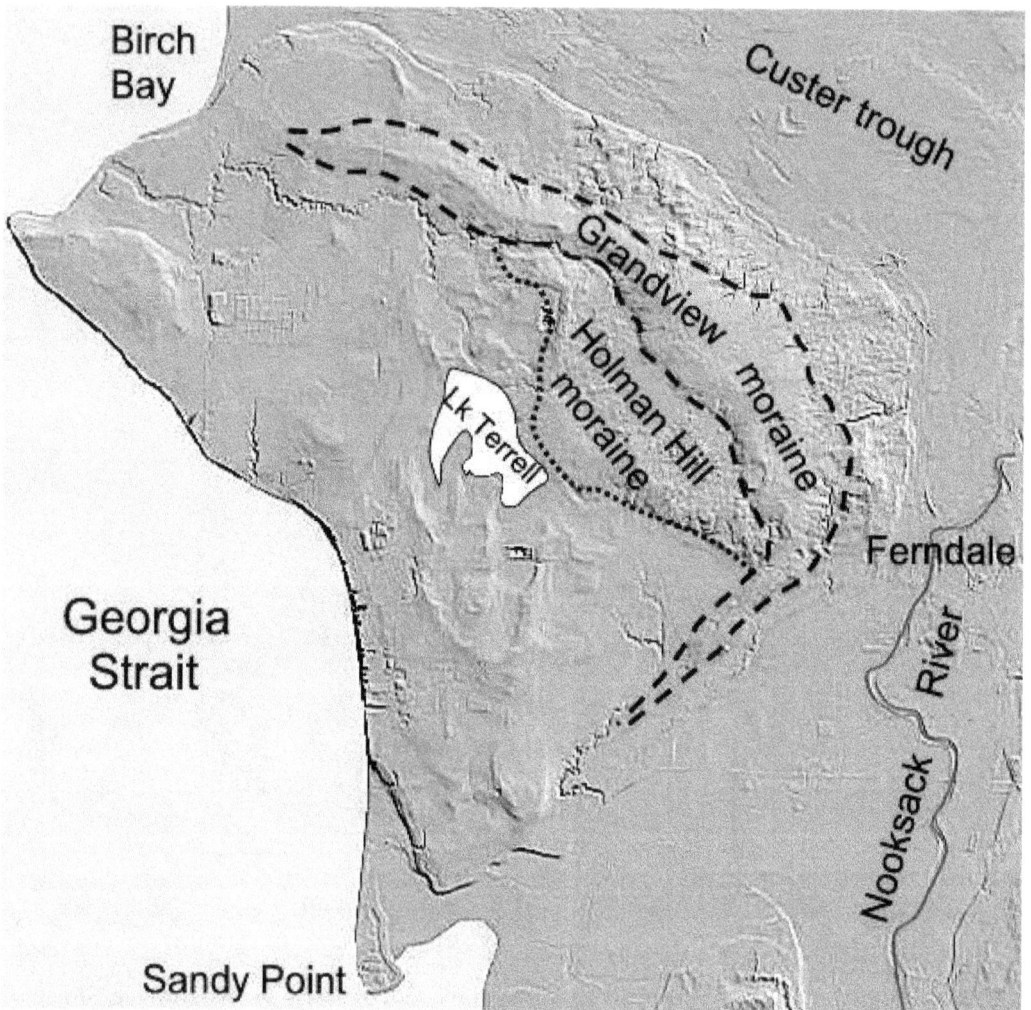

Figure 201. Outermost moraines of the Sumas readvance, locally known as the Holman Hill and Grandview moraines. The radiocarbon age of the moraine is 11,400–11,700years. (Modified from USGS lidar image provided by Whatcom County)

The glacier terminus fluctuated on several occasions as it melted back from its outermost extent. It built at least nine end moraines across various parts of the lowland as it receded toward its source near Sumas. The most prominent moraines were the outermost moraines (Blaine, Grandview, Lummi), the Squalicum moraines (Fig. 202, 203), the Tenmile moraine (Fig. 204), and the Clearbrook moraine near Sumas (Fig. 205).

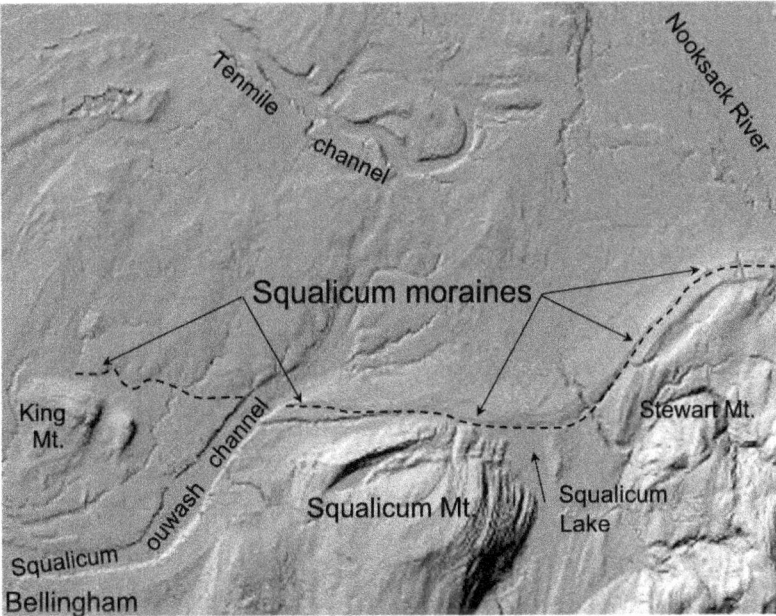

Figure 202. The Squalicum moraines loop across the area between King Mt. and the Nooksack River. The moraines mark the margin of the Cordilleran Ice Sheet remnant about 11,400 radiocarbon years ago. (Modified from USGS lidar image provided by Whatcom County)

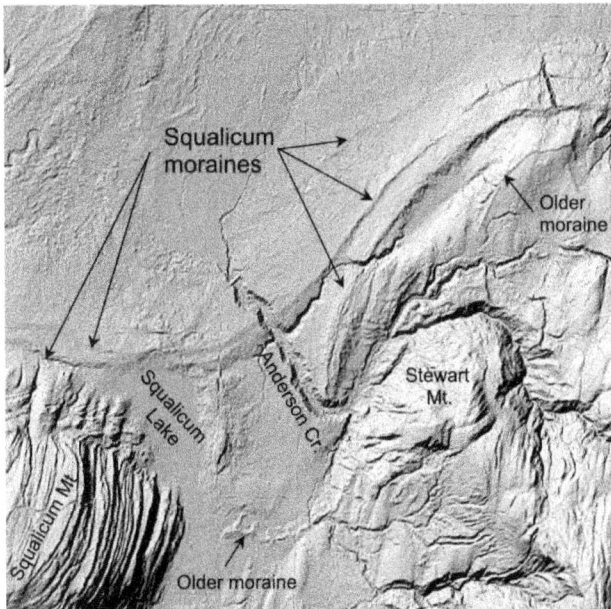

Figure 203. Squalicum Lake moraines north of Squalicum Mt. (Modified from USGS lidar image provided by Whatcom County)

As the ice sheet retreated to the northeast from its maximum extent during the Sumas Stade, it paused long enough to build several parallel moraines just north of Squalicum Lake that extend westward to King Mt. (Fig. 202, 203). The glacier then retreated to the vicinity of Tenmile Creek where it built another moraine just north of the Tenmile Creek channel (Fig. 204). Tenmile Creek channel is incised about 100 feet into Bellingham glaciomarine stony clay and is floored with peat in the NE part of the channel. Wood in basal peat in the channel has been dated at 11,000 radiocarbon years, indicating that meltwater had ceased to flow in the channel by that time because the glacier to the north had retreated northward from the area. Basal peat at the base of bog surrounding Fazon Lake has been dated at 10,400 radiocarbon years, which means that ice had retreated from there by that time.

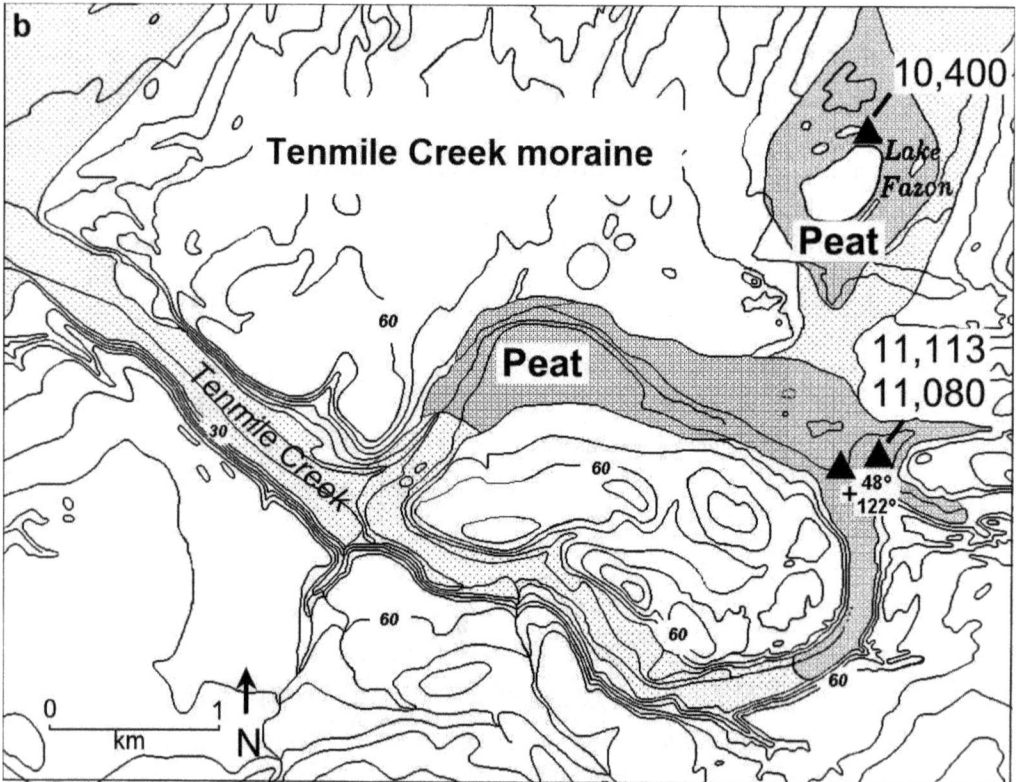

Figure 204. The Tenmile Creek moraine and outwash channel. Numbers are radiocarbon ages of wood from basal peat bogs, indicating that the channel had been abandoned by then.

The youngest of the Sumas Stade moraines was deposited from a glacier lobe SW of Sumas (Fig. 205). The moraine originally extended all the way across the valley but was eroded away by a large outburst flood so only remnants of the moraine are left at the southern and western margins. An outwash plain slopes away from the moraine remnant at the south margin and the Wiser Lake outwash plain was deposited by meltwater from this glacier .

The moraine consists of a hummocky ridge NE of Everson (Fig. 205, 206). An outwash fan (Fig. 205) from this ice was built westward from the vicinity of the town of Sumas. Basal peat from the kettle (Fig. 205) that made Pangborn bog in the outwash fan has been dated at $10,265 \pm 65$ and $10,245 \pm 90$ radiocarbon years indicating that the ice had disappeared by that time.

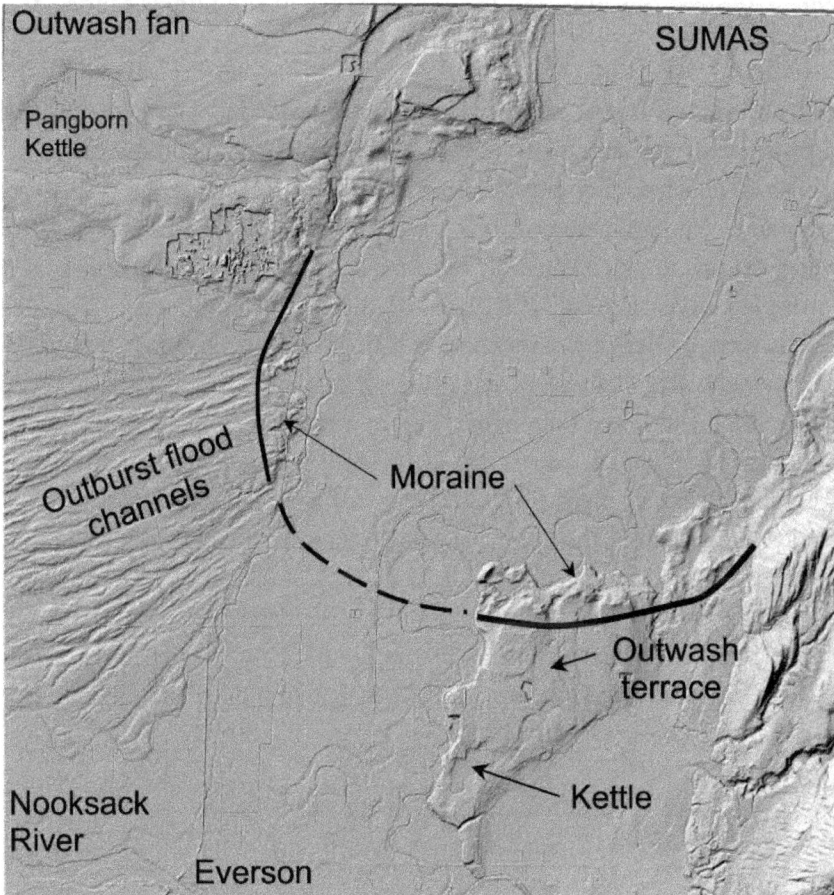

Figure 205. Youngest of the Sumas Stade moraines SW of Sumas. (Modified from USGS lidar image provided by Whatcom County)

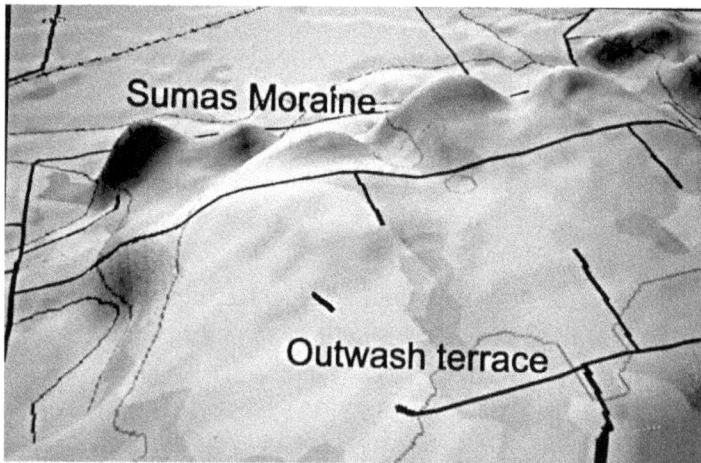

Figure 206. Youngest moraine and outwash terrace of the Sumas Stade SW of Sumas.

A broad area in the north central part of the lowland consists of outwash plains deposited from meltwater streams (Fig. 207). The broad Lynden outwash plain north of Lynden consists of sand and gravel deposited from an ice terminus just across the Canadian border. The SW slope of the outwash plain indicates the direction of flow of the meltwater streams that deposited the outwash sand and gravel that makes up the plain. The outwash plain consists of several levels of slightly different age. The oldest part is near the headwaters of Bertrand Creek in the NW corner of the area. Peat in the bottom of a bog in a meltwater channel of the outwash plain has been dated at 11,413 ± 75 radiocarbon years. Peat in the bottom of a bog in a meltwater channel in a slightly lower terrace of the Lynden plain has been dated at 10,980 ± 250 radiocarbon years.

The Pangborn outwash plain (Fig. 207) west of Sumas slopes westward, indicating that meltwater streams from the youngest Sumas moraine flowed in that direction. The smooth surface of the outwash plain is broken by Pangborn Lake in a deep kettle, a depression made by partial burying of a mass of ice that later melted away, leaving a hole where it had been. Peat from the base of the bog in the kettle has been dated at 10,245 ± 90 and 10,265 ± 65 radiocarbon years ago, indicating that the ice had melted by that time.

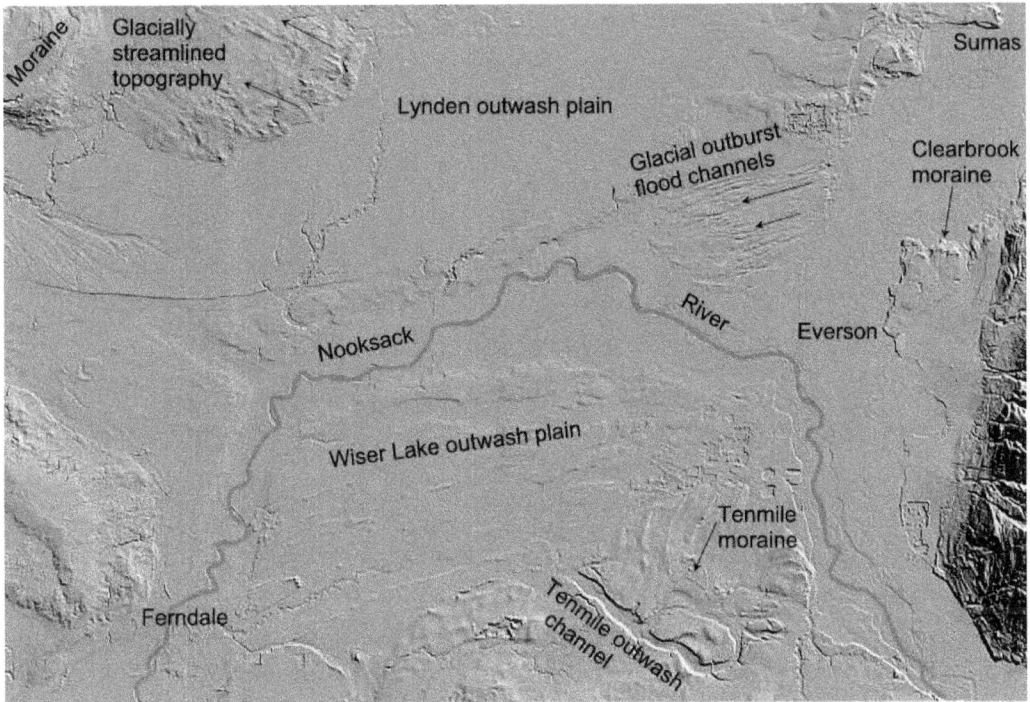

Figure 207. Glacial features of the NW part of the Whatcom County lowland. (Modified from USGS lidar image provided by Whatcom County)

The Wiser Lake outwash plain slopes southwestward from Everson to Ferndale. Several deep channels have been incised into the plain and are filled with peat to depths as much as 35 feet. The outwash was deposited by meltwater streams originating from the youngest moraine near Sumas. The age of the outwash is about 10,000 radiocarbon years.

The Whatcom County landscape is littered with erratic boulders of granite and a distinctive conglomerate. The remnant of the Cordilleran Ice Sheet that deposited the Sumas Stade moraines was a piedmont lobe of ice that issued from the Fraser Valley of British Columbia so the composition of rock debris carried by the glacier was determined by the composition of the bedrock derived from the source area in the Fraser Valley. About one third of the rocks deposited by the Cordilleran Ice Sheet in Whatcom County consists of granite, all of which must have come from British Columbia because no local granite exists. Whatcom County is also strewn with erratic boulders of the distinctive Jackass Mt. conglomerate from the Fraser Canyon (Figs. 208, 209). The conglomerate consists of very tightly cemented cobbles and pebbles surrounded by sandstone.

Figure 208. Huge erratic boulder of Jackass Mt. conglomerate just across the Canadian border near Aldergrove. Boulders of this rock were strewn across the county by Sumas Stade glaciers.

Figure 209. Glacial erratic boulder of Jackass Mt. conglomerate from the Fraser Canyon, B.C. dropped at Donovan St. in Bellingham by Sumas Stade ice about 11,000 years ago. This boulder was originally huge, but was blasted into smaller pieces by the Washington Highway Dept. during freeway construction.

GLACIAL OUTBURST FLOODS—THE RIDDLE OF STRANGE LINEAR TOPOGRAPHY

When I first mapped the geology of the Whatcom County lowland in the 1960s, the rather strange linear topography north of Everson (Fig. 210) was a real puzzle—it was quite different than linear, streamlined topography made by overriding glaciers and no other surface process seemed capable of producing such features. The topography consists of long, low ridges made of silt and clay separated by straight shallow channels that branch and come together. This kind of landform is known as fluted topography. My tentative conclusion at that time was that they must have been formed by some kind of short term flood that swept over the area and created the shallow channels. However, such features were rare and the matter remained without an unequivocal conclusion until just recently when we obtained lidar imagery of the county, which gave a much more detailed view of the topography than could be obtained from ground or air photo observation and other significant features came to light elsewhere in the county.

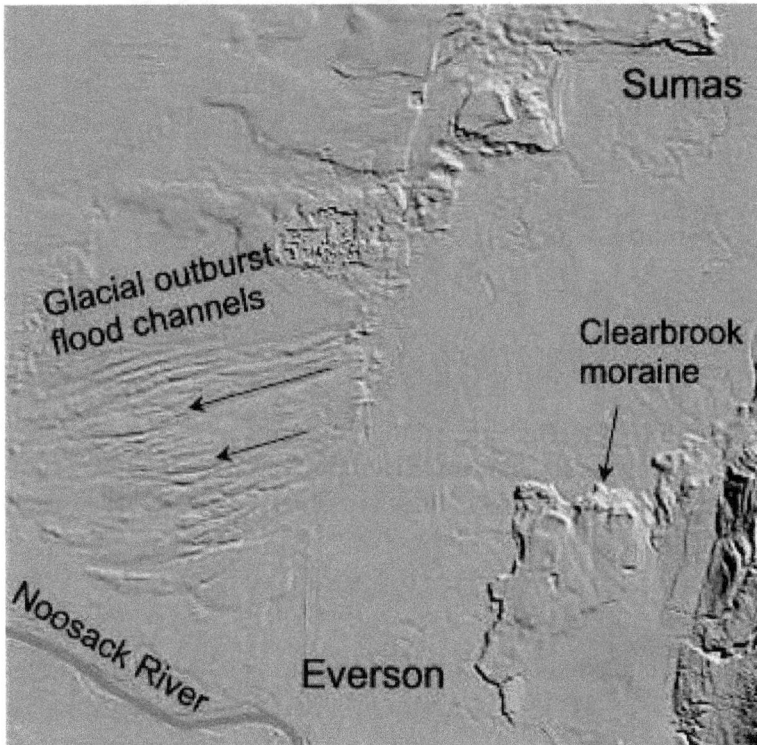

Figure 210. Linear topography made by a glacial outburst flood. (Modified from USGS lidar image provided by Whatcom County)

Figure 211. Streamlined topography (fluted) made by a glacial outburst flood in valleys leading to Drayton Harbor and Birch Bay south of Blaine. Arrows indicate direction of water flow. (Modified from USGS lidar image provided by Whatcom County)

Similar fluted topography occurs in the valleys leading to Drayton Harbor and Birch Bay south of Blaine (Fig. 211). Two levels of linear streamlined topography are visible: (1) the upland area between the valley leading to Drayton Harbor and Birch Bay, and (2) the lower valleys (Fig. 211). Another area of fluted topography occurs at Tennant Lake just south of Ferndale (Fig. 212).

202

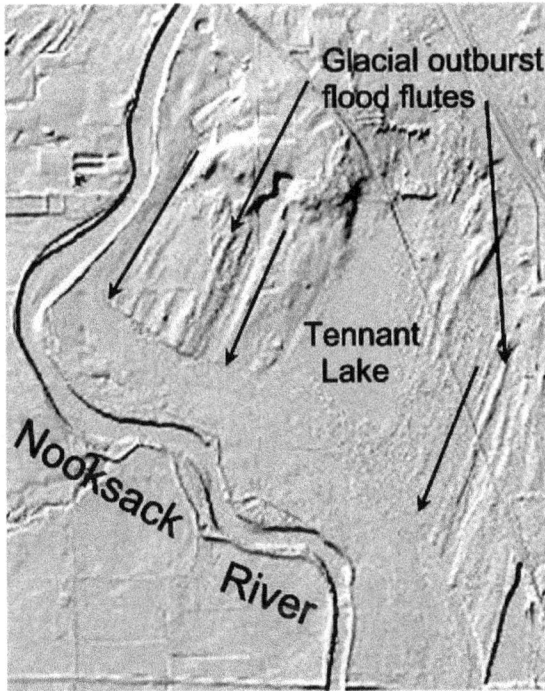

Figure 212. Streamlined topography (fluted) at Tennant Lake south of Ferndale. Arrows indicate direction of water flow. (Modified from USGS lidar image provided by Whatcom County)

The linear, streamlined topography in all three of these areas was made by a glacial outburst flood when an ice-dammed lake, probably in the Chilliwack Valley NE of Sumas, suddenly drained and sent a large volume of water over the area.

Outwash channels

During the last phases of glaciation by the Cordilleran Ice Sheet, meltwater streams cut deep outwash channels in several places in Whatcom County. The most notable are the Squalicum channel (Fig. 213), the Tenmile channel (Fig 213), and several channels in the Wiser Lake outwash plain (Fig. 214). The Squalicum channel was incised by meltwater coming from a glacier margin just south of the Tenmile channel (Fig. 213). The meltwater stream flowed into Bellingham Bay at Bellingham where it built a delta when relative sea level was about 80 feet higher than today. Much of the city of Bellingham is built on this delta.

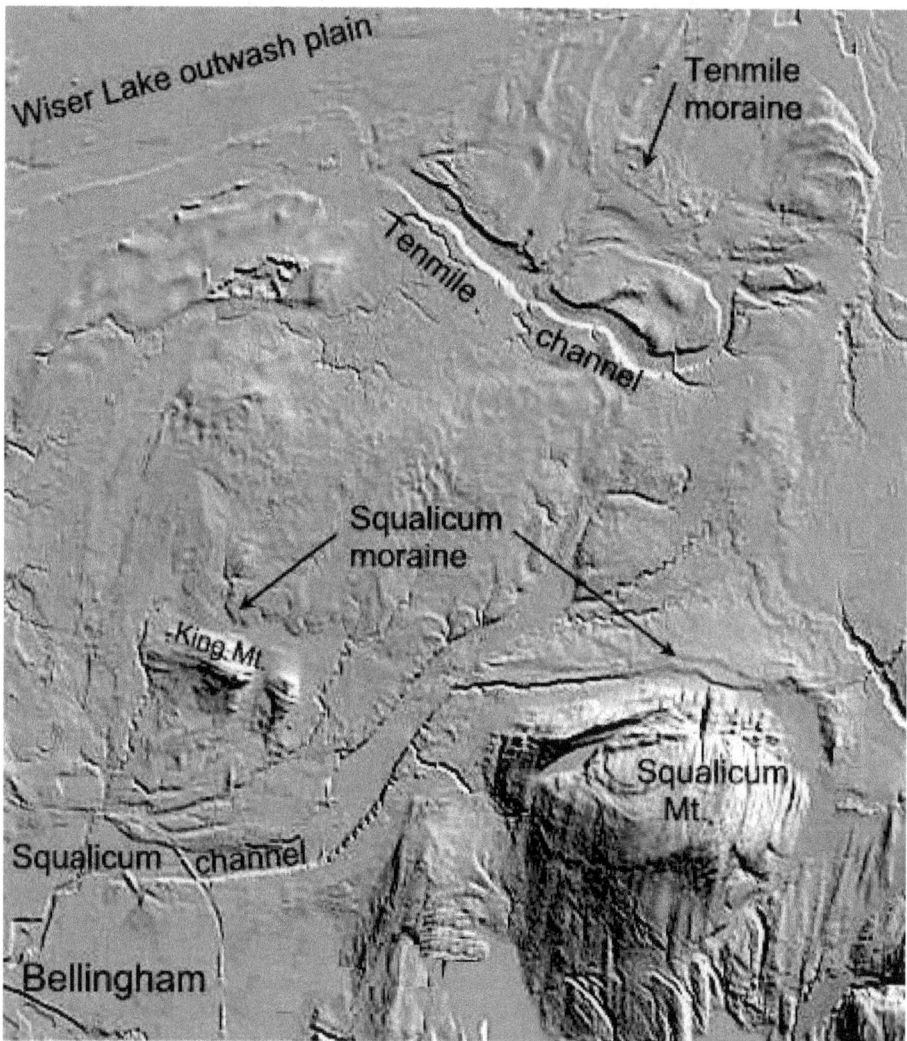

Figure 213. Squalicum and Tenmile outwash channels cut by glacier meltwater streams during the Sumas Stade. (Modified from USGS lidar image provided by Whatcom County)

The Tenmile channel (Fig. 213) was incised into Bellingham glaciomarine stony clay, in part by meltwater from a glacier terminus just to the north of the channel and in part by water from the ancestral Nooksack River, which was diverted through the Tenmile channel by glacial ice blocking the normal Nooksack channel south of Everson.

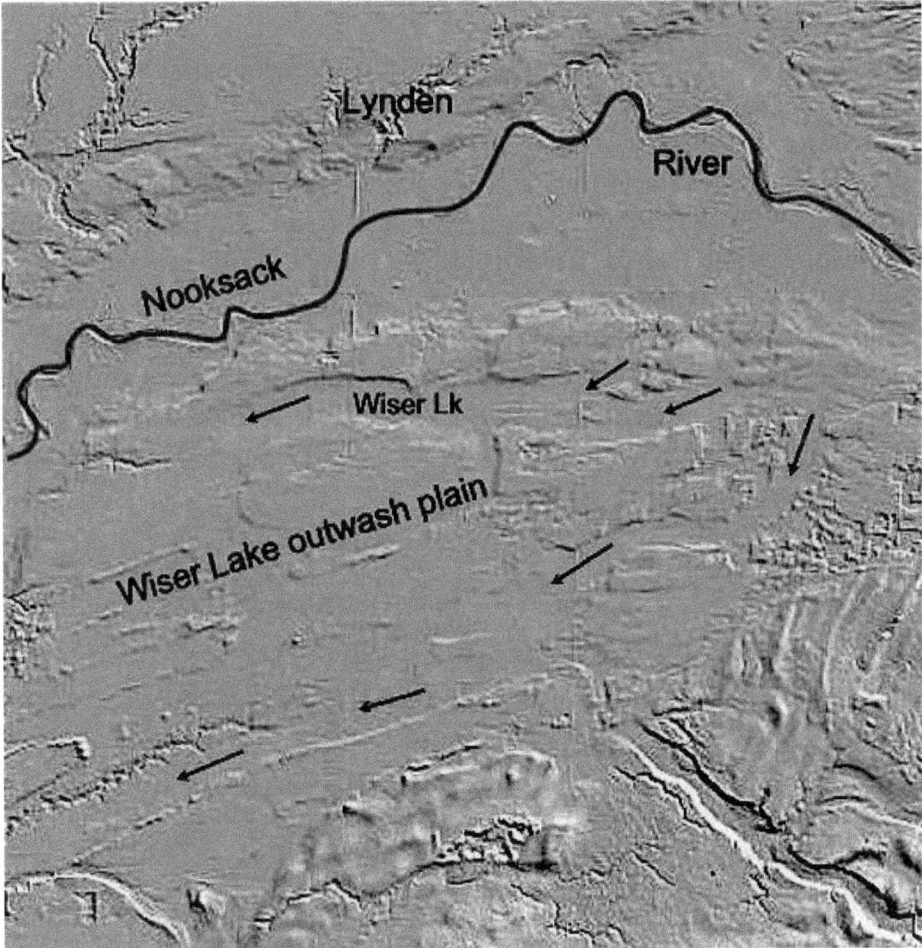

Figure 214. Meltwater channels in the Wiser Lake outwash plain. Arrows point in the direction of meltwater flow. (Modified from USGS lidar image provided by Whatcom County)

Meltwater channels incised into the Wiser Lake outwash plain (Fig. 214) are filled with peat, in places up to 35 feet thick. Thus, although their topographic expression is fairly subdued, their true depth is considerable.

205

THE NOOKSACK ALPINE GLACIER SYSTEM

ALPINE GLACIERS IN THE NOOKSACK VALLEY FOLLOWING RETREAT OF THE CORDILLERAN ICE SHEET

At the maximum advance of the Cordilleran Ice Sheet during the last Ice Age about 17,000 years ago, only peaks in the North Cascades higher than 6,000 feet stood above the surface of the ice. The ice flowed essentially north–south, as shown by glacial grooves and striations near Mt. Baker. Rapid melting of the glacier between 17,000 and 14,000 years ago resulted in lowering of the ice surface below ridge crests in the Nooksack drainage and glacial activity thereafter became topographically controlled by ridges and valleys. The glaciers were no longer connected to the Cordilleran Ice Sheet and valley glaciers in the upper Nooksack Valley were fed by glaciers on Mt. Baker, Mt. Shuksan, and the Twin Sisters Range (Figs. 215–217). Remnants of the Cordilleran Ice Sheet persisted in the lowland at that time but were separated from the Nooksack Valley glaciers by several 4,000–foot ridges, higher than the surface of the ice sheet.

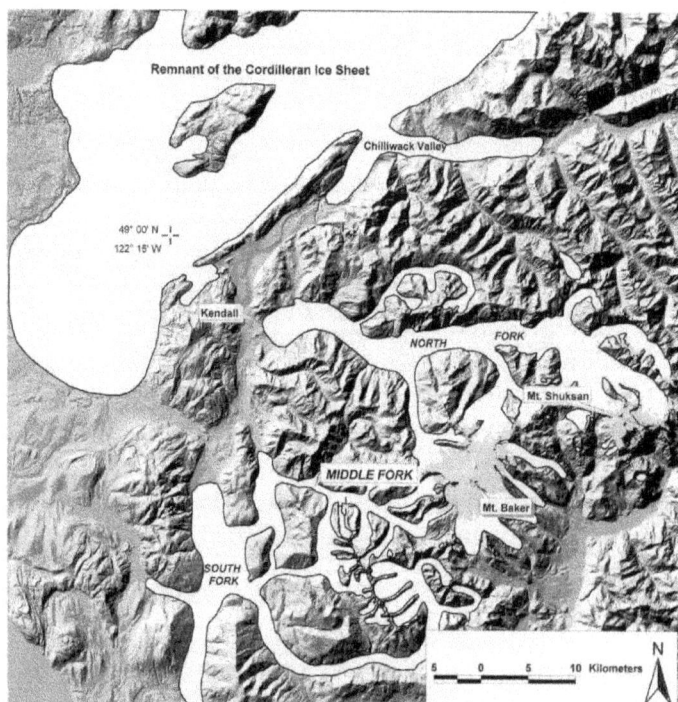

Figure 215. The Nooksack Alpine Glacier System about 12,000 years ago.
(From Kovanen and Easterbrook, 2001)

206

These alpine glaciers, known as the Nooksack Alpine Glacier System (NAGS), deposited sediments in the Middle, North, and South forks of the Nooksack drainage, 15–25 miles down-valley from their sources. The glacier split into three topographically controlled alpine glaciers (Fig. 216): (1) the South Fork glacier terminated near Cranberry Lake north of Sedro Woolley, (2) the Middle Fork glacier reached the southeast arm of Lake Whatcom, and (3) the North Fork glacier extended from Mt. Shuksan to Kendall.

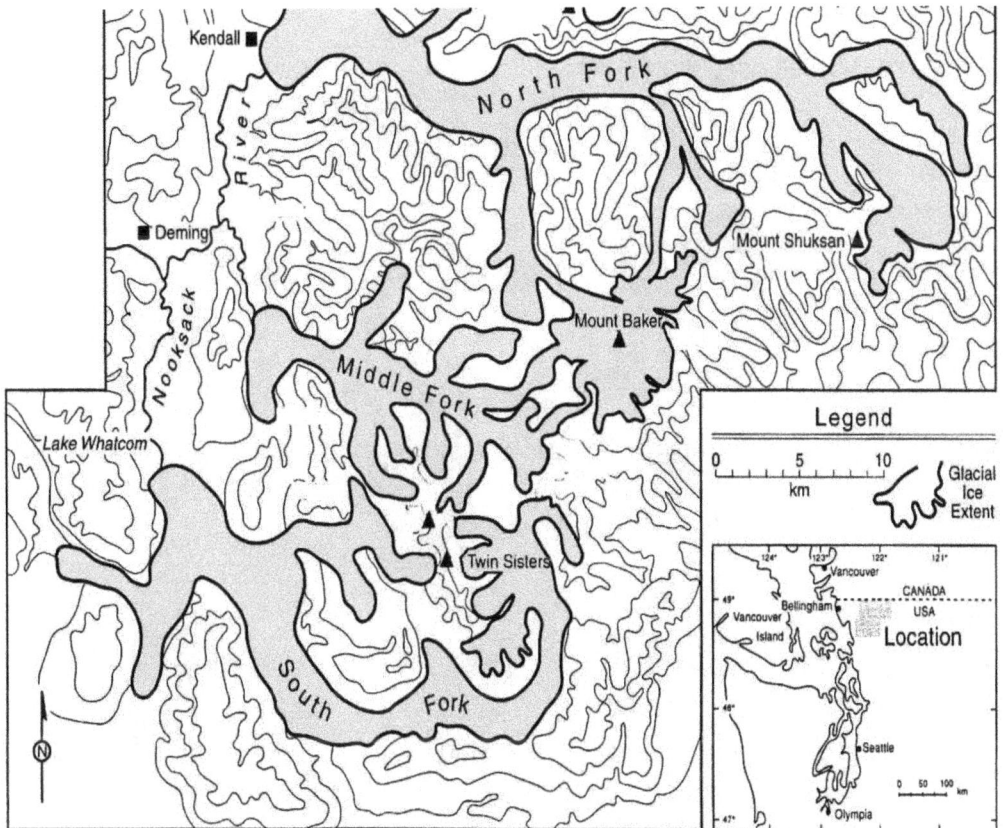

Figure 216. The Nooksack Alpine Glacier System. Shaded areas are glaciers.

Middle Fork alpine Ice Age glaciers

The Mosquito Lake valley, a tributary valley of the Nooksack Middle Fork with no stream in it, is filled with variously shaped mounds and hollows composed of sand and gravel (Fig. 218) deposited from a stagnant glacier. Many of the depressions are up to 150 feet deep and are filled with water. From the ground, the topography appears to be randomly spaced mounds and depressions, but aerial photographs reveal that the mounds form a series of undulating, concentric ridges in the northern part of the valley. In the southern part of the valley, the deep hollows and small irregular hills indicate deposition from a stagnating glacier.

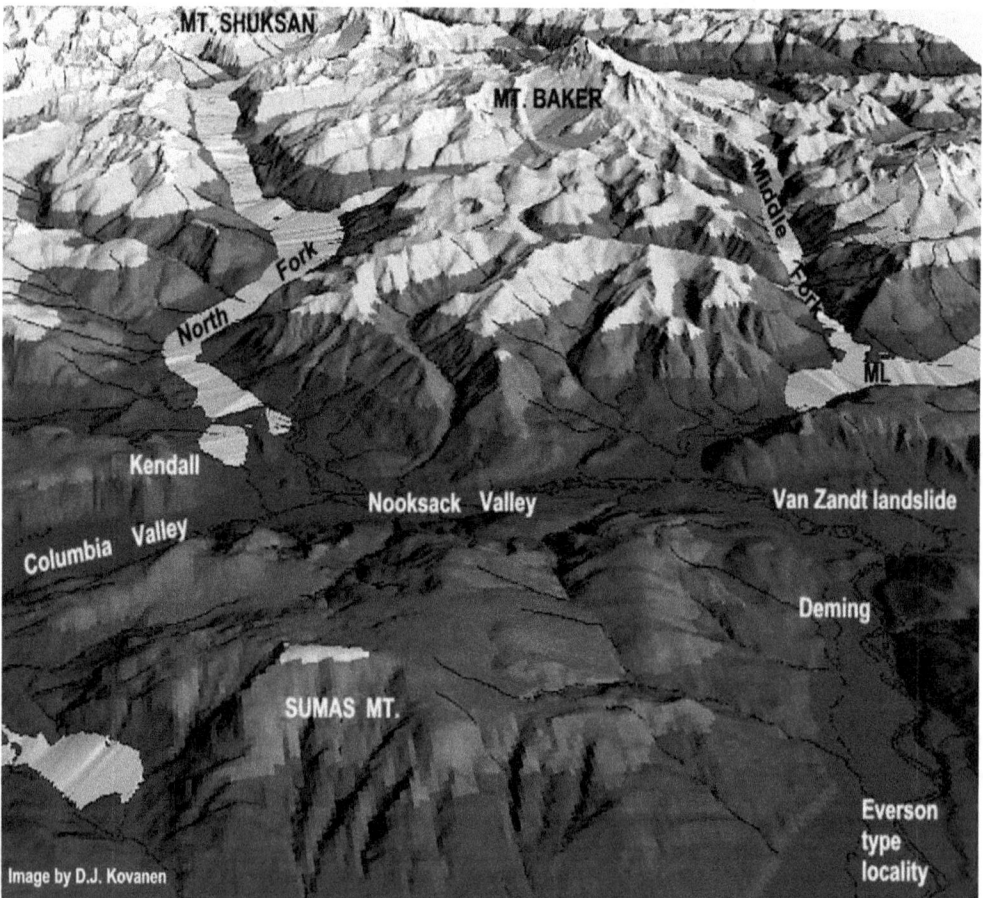

Figure 217. Reconstruction of the Nooksack Alpine Glacier System (NAGS) in the Nooksack drainage. Long valley glaciers extended down the North Fork to Kendall and down the Middle Fork to Mosquito Lake. (From Easterbrook, 2003)

208

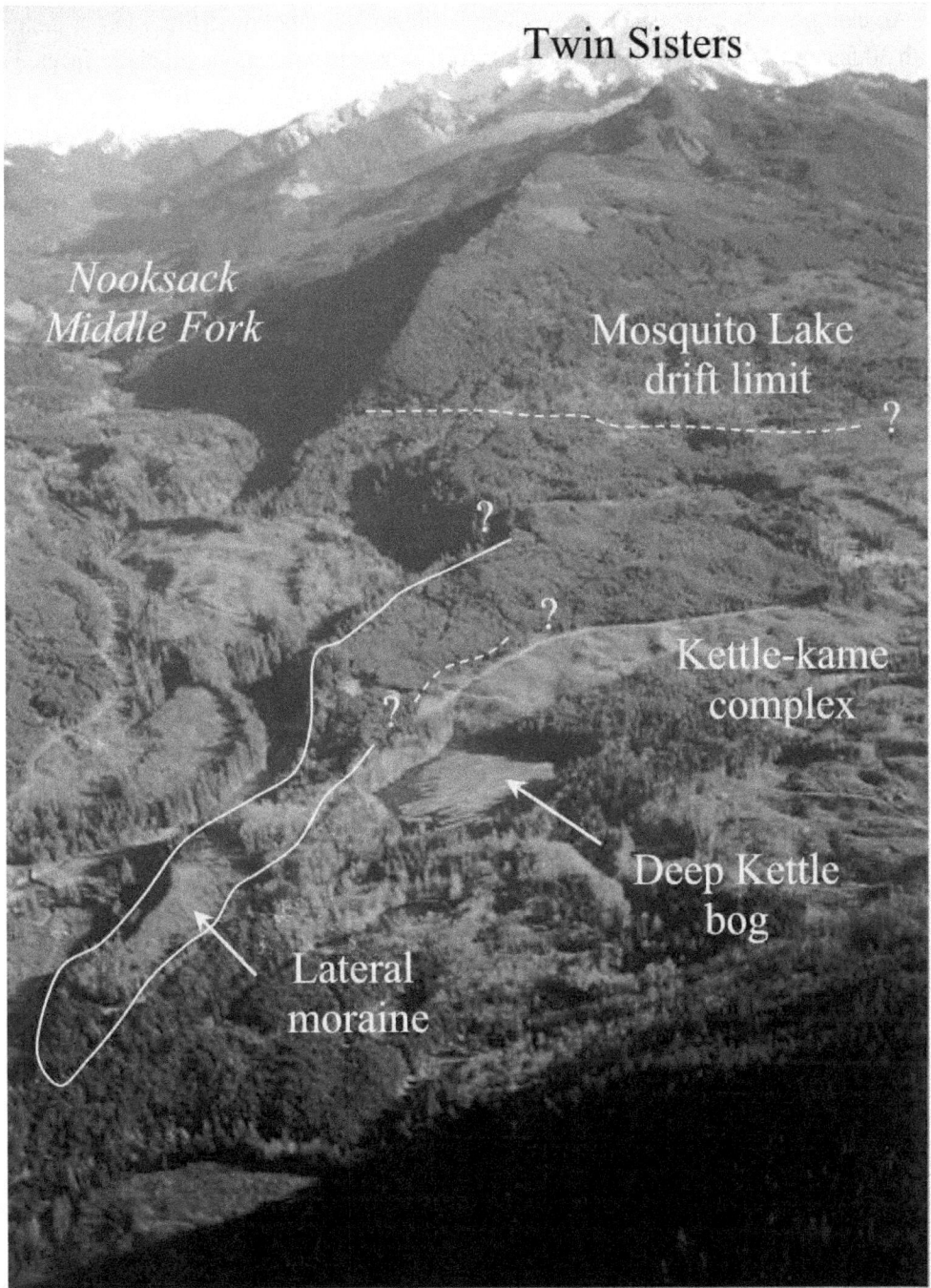

Figure 218. A long ridge (lateral moraine) draped over stagnant ice deposits (kettle-kame complex) and deep hollows filled with peat near Mosquito Lake. These deposits are slightly older than 12,350 radiocarbon years.

An elongate ridge drapes across the irregular mounds and hollows (Fig. 218). It is capped with 30 feet of poorly sorted glacial deposits (till) containing numerous glacially faceted and striated boulders from Mt. Baker and the Twin Sisters Range (Fig. 219). The glacial till is underlain by up to 60 feet of sand and gravel displaying faults and other collapse structures suggesting deposition against stagnant ice. Erratic boulders more than three feet in diameter that litter the moraine surface consist mostly of Mt. Baker lava (82.6%), Twin Sisters dunite (8.6%), Darrington phyllite, and Chuckanut sandstone, indicating a local source.

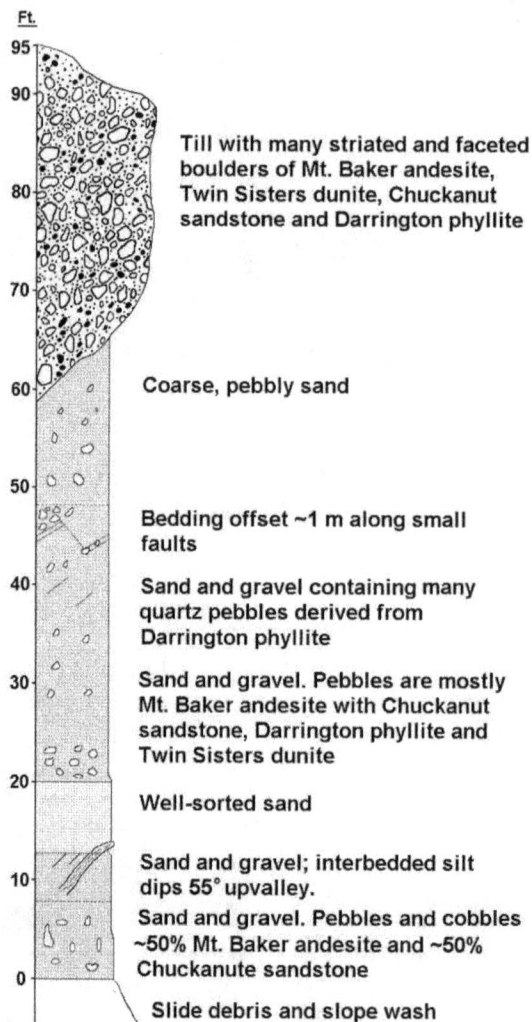

Figure 219. Sediments composing the long lateral moraine near Mosquito Lake. The composition of stones in the glacial till indicate that they were derived upvalley from local bedrock, rather than from the Cordilleran Ice Sheet.

210

Figure 220. Bouldery glacial till (top) making up the lateral moraine near Mosquito Lake. The boulders and cobbles in the till are mostly Twin Sisters dunite and Mt. Baker lava, Chuckanut sandstone, and Darrington phyllite, indicating a local source upvalley, rather than the Cordilleran Ice Sheet.

Two 65–foot–deep sediment cores were obtained from a deep peat bog north of Mosquito Lake. At the base of the core, rootlets in gravel were dated at 12,356 ± 115 radiocarbon years and wood on the gravel was dated at 12,165 ± 95 radiocarbon years. These dates indicate that the ice that once occupied the depression had melted away before 12,350 radiocarbon years ago.

The Middle Fork alpine glacier retreated eight miles up-valley from Mosquito Lake where it built a lateral moraine during the Younger Dryas cold period at the end of the last Ice Age that buried logs dated at 10,680 ± 70 and 10,500 ± 70 radiocarbon years ago.

211

North Fork Ice Age alpine glaciers

The Nooksack North Fork originates at a large glacial basin on the NW flank of Mt. Shuksan and extends westward in a long, broad, glacial trough. Twenty five miles downvalley from Mt. Shuksan, remnants of two moraines are preserved along the north side of the valley (Fig. 221). The outermost of the two moraines, the Kendall moraine, is a distinct ridge several hundred yards long. The Maple Falls moraine lies a mile upvalley.

Figure 221. Moraines in the Nooksack North Fork near Kendall and Columbia Valley, a former course of the Chilliwack River in British Columbia.

The composition of stones in the Kendall moraine shows a distinct, local, upvalley source. In contrast to deposits of the Cordilleran Ice Sheet, which consists mostly of granite, the Kendall moraine is composed mostly of Chuckanut sandstone (54%) and Mt. Baker lava (12%). The Mt. Baker lava is significant because it must have come from Glacier Creek, a tributary of the upper Nooksack near the town of Glacier. The moraine also contains cobbles from a unique outcrop of Paleozoic volcanic rock found only in the upper North Fork drainage.

The composition of stones in the glacial till means that ice must have flowed northward from Mt. Baker into the North Fork valley in the direction opposite to the earlier flow of the Cordilleran Ice Sheet, confirming that the Kendall moraine was built by a local alpine glacier rather than the ice sheet.

An unusually large number of stones in the Kendall moraine are glacially faceted, striated, and polished, many more than at other localities in the Pacific Northwest. The reason for so many glacially abraded stones appears to be the long transport distance from the glacial source 25 miles upvalley.

The town of Maple Falls is built on a terrace above the Nooksack River (Fig. 222). The terrace is an old alluvial fan built by a meltwater stream coming from a glacier at Silver Lake about four miles north of Maple Falls where three moraines hold in Silver Lake (Fig. 223).

Columbia Valley north of Kendall (Figs. 222, 224-226) is more than 10 miles long and up to two miles wide, but is not occupied by a stream. Large abandoned channels and meander scars that mark the valley floor and sides were clearly made by a stream much larger than the minor, local creeks now in the valley. Just across the Canadian border the valley floor drops abruptly more than 400 feet in a steep bluff to a much lower valley floor at Cultus Lake (Fig. 224–226). This interesting situation poses the question–what happened to the stream that made Columbia Valley? The answer is quite clear upon looking across the Canadian border. The valley continues for many miles upvalley into British Columbia where it is occupied by the Chilliwack River. At one time, the Chilliwack River must have flowed southward as a tributary to the Nooksack drainage, joining the Nooksack North Fork at Kendall. When the Cordilleran Ice Sheet melted down below the level of ridge tops in the North Cascades, a remnant of ice remained in the Chilliwack Valley of British Columbia just north of the border and meltwater flowed southward down Columbia Valley. Meltwater sand and gravel more than 400 feet thick was deposited in Columbia Valley and was banked against the ice terminus at Cultus Lake. When the ice finally melted away, a 400 foot high bluff was left where the ice had been and Cultus Lake was formed (Figs. 225, 226),. Because of the high bluff, the Chilliwack River could no longer flow southward down Columbia Valley so the river spilled over a low divide at Vedder Mt. and cut a new course to the Fraser River.

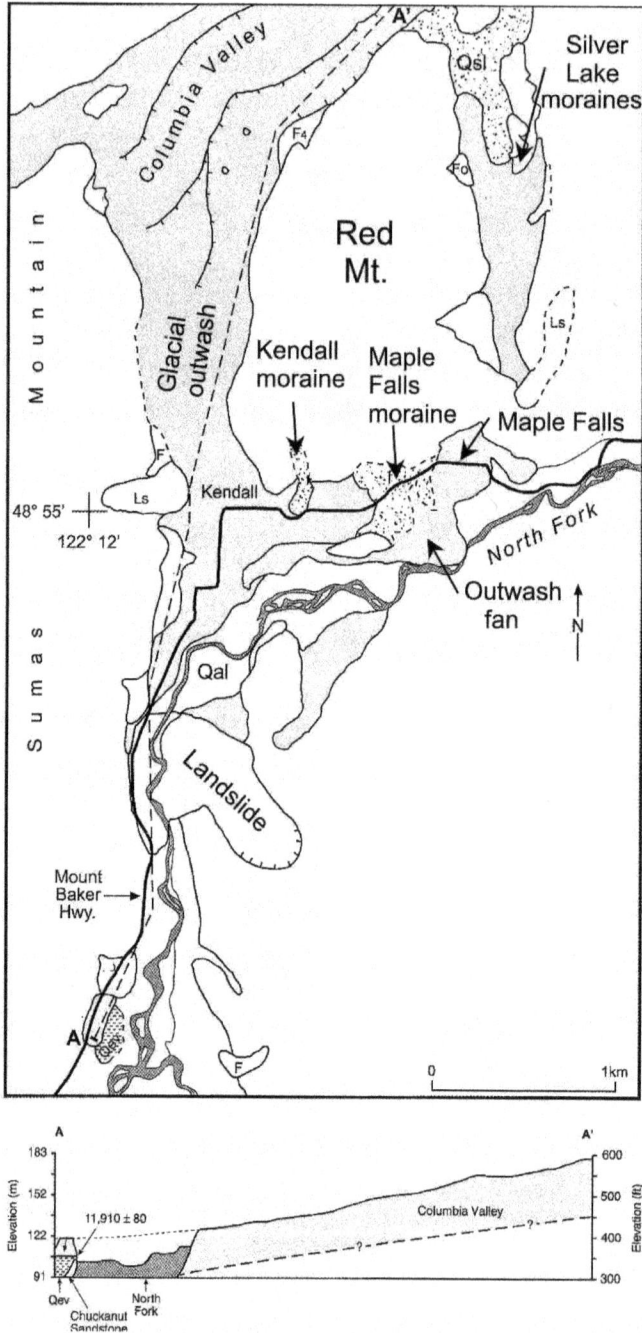

Figure 222. Columbia Valley, Kendall moraine, Maple Falls moraine, and Silver Lake moraines. (Kovanen and Easterbrook, 2001)

Figure 223. Moraines holding in Silver Lake. Meltwater from ice at Silver Lake deposited the large fan upon which Maple Falls is built.

Wood from the bluff at Cultus Lake has been dated at 11,300 ± 100 radiocarbon years, which is close to the age of the meltwater sand and gravel in Columbia Valley. The surface of outwash terraces from alpine moraines in the Nooksack North Fork merges smoothly with Columbia Valley outwash near Kendall, indicating that they were formed at the same time. Two charcoal layers in the sediments of the terraces have been dated at 10,788 ± 77 and 10, 603 ± 69 radiocarbon years (Fig. 227). The terrace sediments rest upon glaciomarine stony clay containing wood dated at 11,910 ± 80 radiocarbon years. Thus, outwash terraces of the North Fork and the outwash fill in Columbia Valley appear to be 10,700 radiocarbon years old.

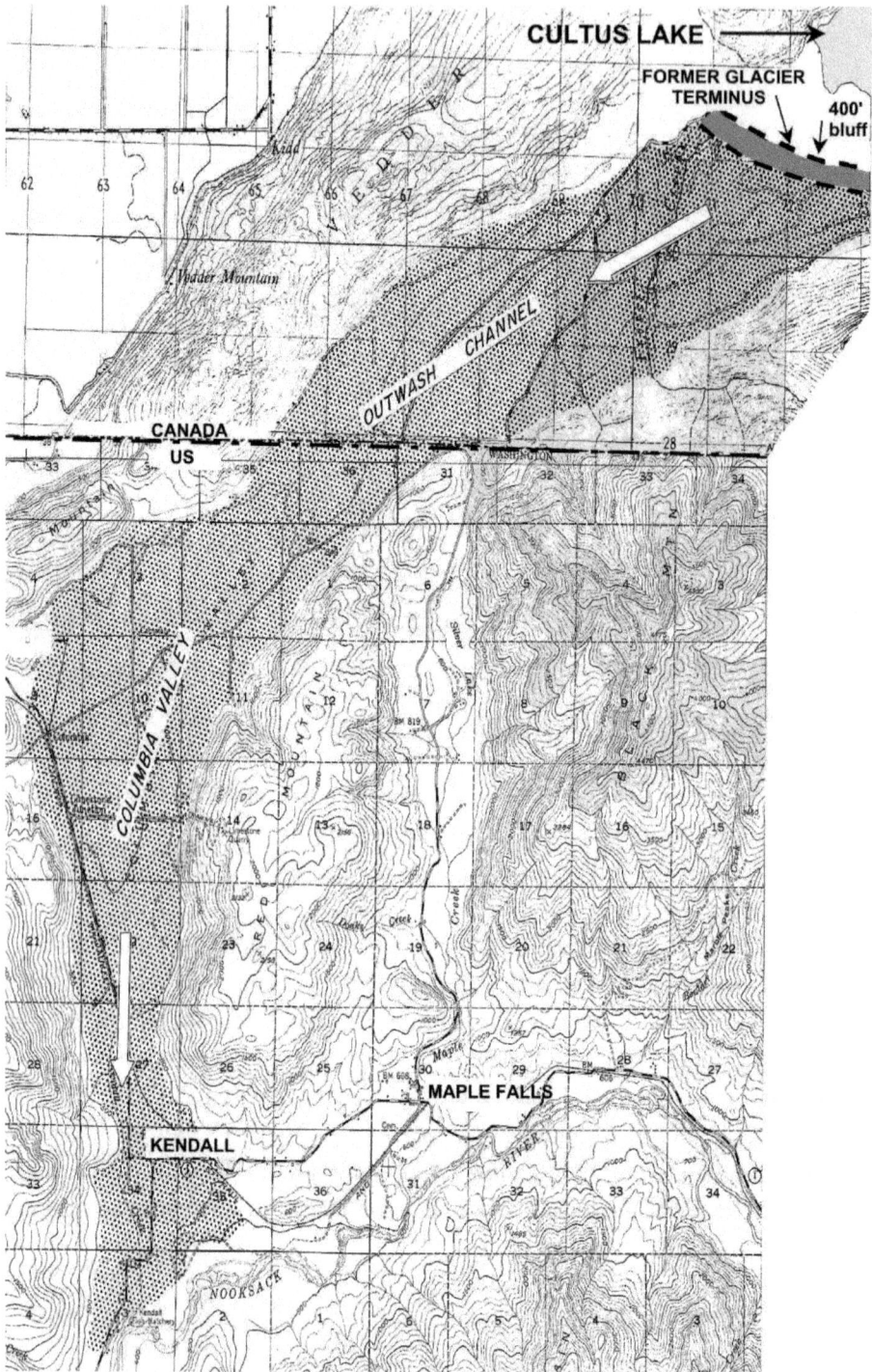

Figure 224. Map of Columbia Valley outwash from ice at Cultus Lake.

Figure 225. Columbia Valley outwash from ice at Cultus Lake.

Figure 226. Upper end of Columbia Valley overlooking Cultus Lake and the Chilliwack Valley. The glacier was banked against the ice-contact face when Columbia Valley was filled with meltwater sand and gravel.

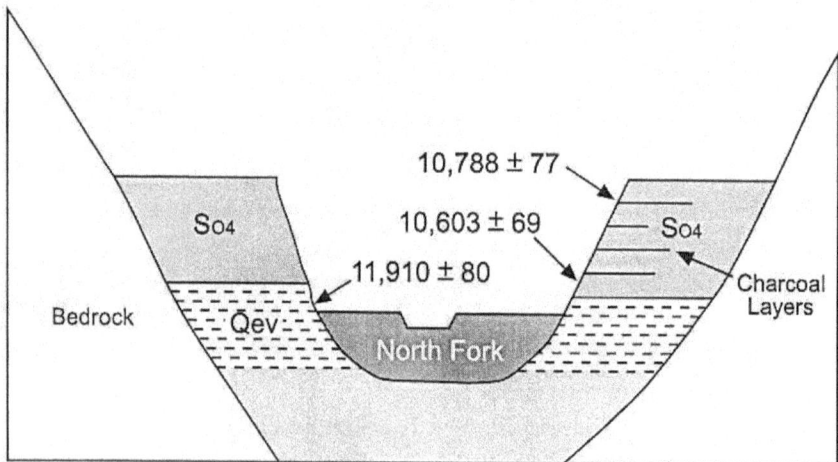

Figure 227. Geologic cross section of the North Fork Valley showing the radiocarbon dates of the outwash terraces south of Kendall.

218

South Fork Ice Age alpine glaciers

The South Fork glacier, which extended 25 miles downvalley from the Twin Sisters Range to Lake Whatcom (Fig. 216), deposited a lateral moraine that makes a ridge along the south shore of Lake Whatcom (Fig. 228) and terminates as a peninsula into the lake. The moraine is composed almost entirely of Darrington phyllite from the valley to the east and dunite from the Twin Sister Range.

The moraine contains many large, faceted, and striated dunite boulders derived from the Twin Sisters Range. Of 109 boulders at least two feet in diameter revealed in a deep excavation one mile west of the east end of the lake, 91 were Darrington phyllite and 11 were Twin Sisters dunite, indicating that the glacier that deposited the moraine came from the Twin Sisters Range to the east, rather than from the Cordilleran Ice Sheet.

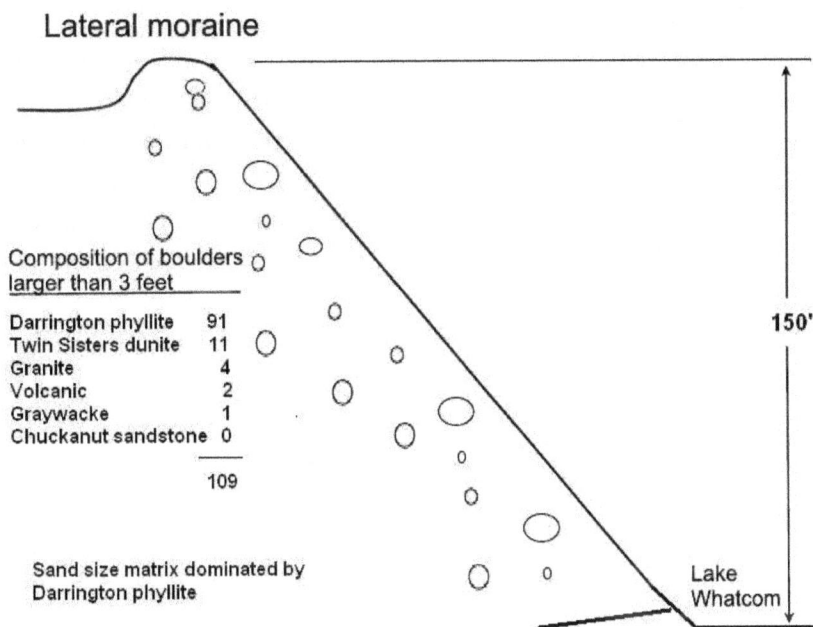

Figure 228. Cross section through the lateral moraine on the south shore of Lake Whatcom one mile west of the east end of the lake. The composition of boulders in the moraine proves that the glacier came from Twin Sisters Range as part of the Nooksack Alpine Glacial System.

A bog about 35 feet deep surrounds Cranberry Lake at the terminus of the former South Fork glacier seven miles south of Wickersham. Basal bog dates (12,215 to 12,733 radiocarbon years), are generally equivalent to those at the deep bog in the Middle Fork valley near Mosquito Lake, suggesting that the glaciers in both valleys were the same age.

Meltwater from the southern lobe of the South Fork glacier drained southward in the South Fork valley, rather than northward as it does now. Meltwater crossed the low divide between the South Fork valley and the upper Samish River valley and flowed down the Samish River valley to the sea in the Skagit Valley just north of Burlington. River meanders in the lower part of the Samish valley are many times larger than those of the present stream (Fig. 200), indicating that the meltwater discharge was substantially larger than that of the present. The meanders are eroded 200 feet into Everson glaciomarine stony clay (11,700 radiocarbon years old) and cut across post-glaciomarine beach ridges at a marine limit of 214 feet before terminating at a marine delta about 100 feet above present sea level (Fig. 200). The age of the South Fork outwash is younger than the glaciomarine stony clay (11,700 radiocarbon years) and younger than the marine shorelines which have been dated at 11,700 ± 110 radiocarbon years. Therefore, the age of the outwash channel must be younger than ~11,700 radiocarbon years.

THE LATE ICE AGE IN THE HIGH CASCADES

HEATHER MEADOWS–ARTISTS POINT

The alpine scenery at Heather Meadows (Fig. 229) is among the most beautiful in the world. From Heather Meadows and Artists Point, a sweeping panorama of alpine peaks greets the eye. To the east is Mt. Shuksan, a 9,127 foot high peak with hanging glaciers that break off in thundering avalanches all summer. To the south is Baker Lake with Glacier Peak on the far skyline. To the southwest is glacier clad Mt. Baker (10,785') with its steaming volcanic crater. To the west is Table Mt, a long ridge of lava that was once a valley, with glacial basins occupied by Bagley and Chain Lakes. Visible to the north are Tomyhoi Peak (7,451'), American Border Peak (8,068'), Red Mt. (also known as Mt. Larrabee) (7,868'), Winchester Mt. (6,521'), Goat Mt. (6,891'), and Mt, Sefrit (7,191').

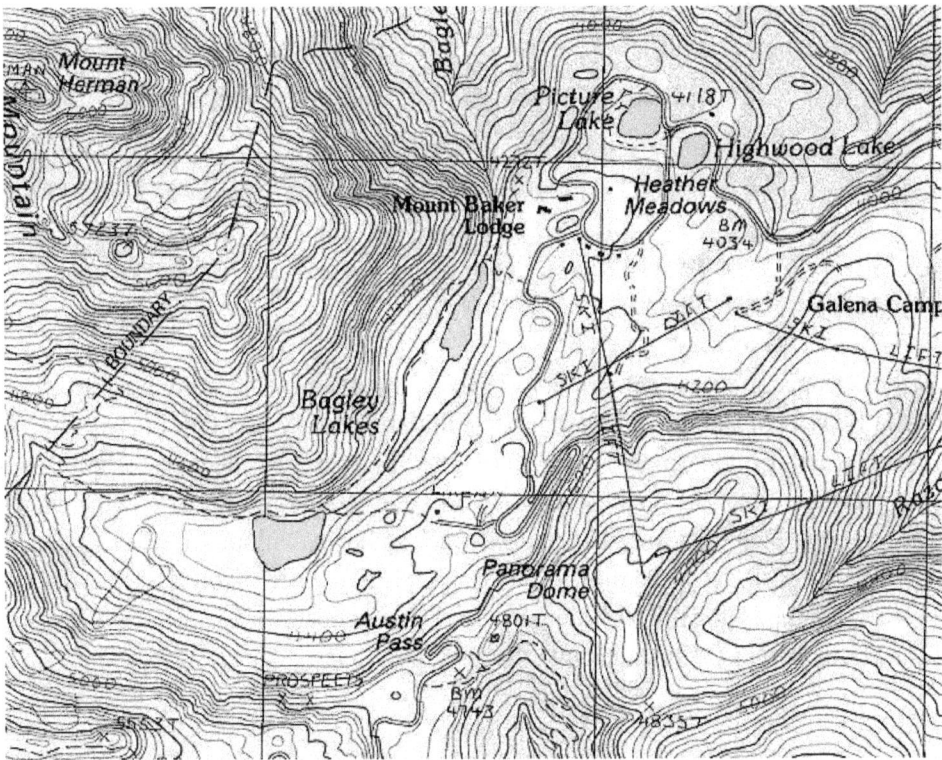

Figure 229. Map of Heather Meadows area. Bagley Lakes, Picture Lake, and Highwood Lake were all made by alpine glaciers. (USGS topographic map)

During the maximum of the last Ice Age (about 17,000 to 10,000 years ago), the Cordilleran Ice Sheet covered the North Cascade Range below elevations of about 6,000 feet. Erratics of Paleozoic rocks from Mt. Herman lie atop Table Mt. at an elevation of 5,742 feet, indicating that the ice sheet picked up the boulder as it rode over Mt. Herman (Fig. 230) to the north and deposited it on top of Table Mt. (Fig. 230) Adding a few hundred feet for ice thickness means that the surface of the Cordilleran Ice Sheet here must have been more than 6,000 feet above sea level here.

Figure 230.   Glacial erratic of Paleozoic Chilliwack volcanic rock (left) carried from Mt. Herman (right) to Table Mt. and dropped on the summit Mt. Baker lava flows by the Cordilleran Ice Sheet.

Well-developed grooves and striations have been carved by the Cordilleran Ice Sheet in the lava at Heather Meadows and on the crest of the ridge at Artists Point (Figs. 231, 232). Significantly, all of the grooves are oriented within a few degrees of north-south, indicating that the direction of ice flow was almost due south down the axis of the range and the ice was not topographically constrained during the maximum Cordilleran Ice Sheet glaciation. This conclusion is confirmed by the complete absence of Mt. Baker andesite and other local rock types in till in the adjacent lowland, indicating that the ice did not flow westward from the North Cascades into the lowland.

Figure 231. North–south glacial grove and polished rock made by scouring of the Cordilleran Ice Sheet at Artists Point.

Figure 232. Lava flow polished by glacial scour of the Cordilleran Ice Sheet at Artists Point.

Picture Lake–Highwood Lake Moraines

The Cordilleran Ice Sheet shrank rapidly between 14,500 and 13,000 radiocarbon years ago and disappeared from the North Cascades by 12,500 radiocarbon years before present, allowing local alpine glaciers to occupy the region. Alpine ice filled the basins now occupied by Bagley Lakes and Highwood Lake (Fig. 229). Picture Lake, with its beautiful reflections of Mt. Shuksan (Fig. 233), is among the most photographed areas in North America.

Figure 233. Moraine holding in Picture Lake, Heather Meadows.

Picture Lake (Fig. 233) and nearby Highwood Lake are held in by moraines mantled with 6,850 year old Mazama ash, which can be seen in the roadcut (Fig. 234) across the road from the Picture Lake trail entrance. The trail circles the lake and offers several view points of the spectacular scenery, which is at its best during the brilliant red fall colors.

224

Fig. 234. 6,850–year–old Mazama ash on the moraine at Picture Lake, Heather Meadows.

A radiocarbon age of 9,410 ± 60 radiocarbon years before present was obtained from basal peat in Highwood Lake, indicating that ice had retreated from the moraine by then and that the moraine must be older than 9,400 radiocarbon years before present.

Bagley and Iceberg Lakes

Bagley Lakes lie on the floor of a glacial cirque (Figs. 235-237) that was eroded by a glacier extending from Table Mt. to Picture Lake (Fig. 229). Iceberg Lake (Fig. 238) on the opposite side of Table Mt. also lies on a cirque floor made by an alpine glacier during the last Ice Age.

Figure 235. Bagley Lakes from Austin Pass. The alpine glacier that made the lakes extended to the far end of the photo.

Figure 236. Upper Bagley Lake in a glacially eroded basin (cirque).

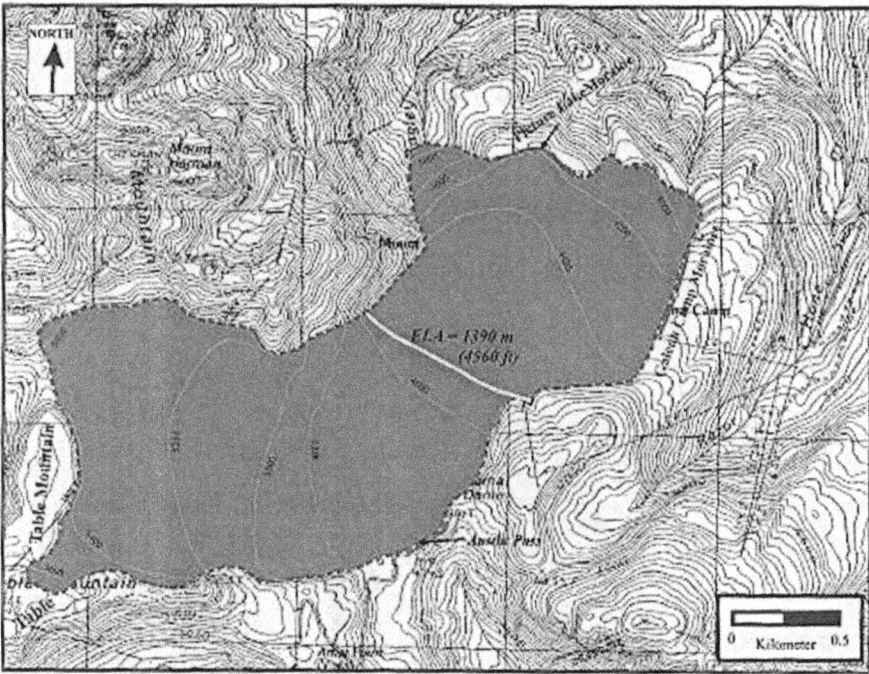

Figure 237. Reconstruction of the late Ice Age alpine glacier in the Bagley Lakes–Picture Lake area. (Modified from Burrows, 2000)

Figure 238. Iceberg Lake, one of the Chain Lakes on the west side of Table Mt. occupies a glacial cirque of the last Ice Age.

227

Other examples of last Ice Age alpine cirques and moraines include Park Butte (Fig. 239) and Pocket Lake (Fig. 240) above Schreibers Meadow on the SW flank of Mt. Baker. Many other glacial basins in the North Cascades contain similar moraines.

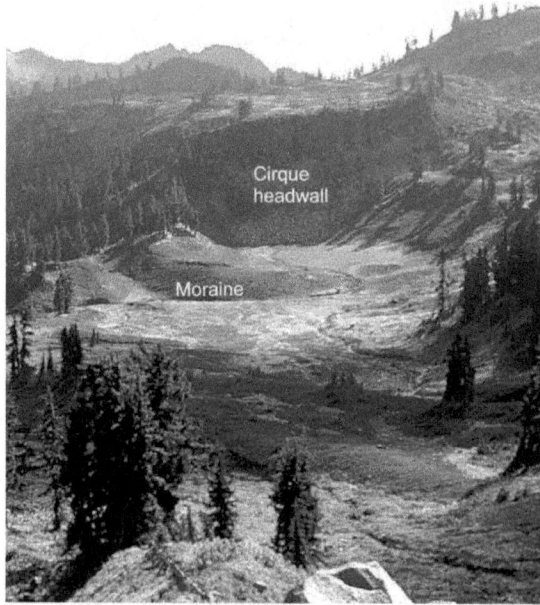

Figure 239. Small glacial basin (cirque) with a moraine, upper meadow above Schreibers Meadow near Park Butte.

Figure 240. Pocket Lake in a glacial basin (cirque) above Schreibers Meadow. The moraine holding in the lake contains wood dated at 8,400 radiocarbon years.

CHAPTER 5

THE GEOLOGIC ORIGIN OF LAKE WHATCOM

Lake Whatcom occupies a narrow NW-SE trending valley about 13 miles long and one mile wide (Figs. 241, 242). The valley sides rise from a lake level of 307 feet to over 2,000 feet. The lake consists of four connected basins. The northern two basins are small and quite shallow, and the southern two are much larger and much deeper. The deepest basin is 328 feet deep and is slightly below sea level.

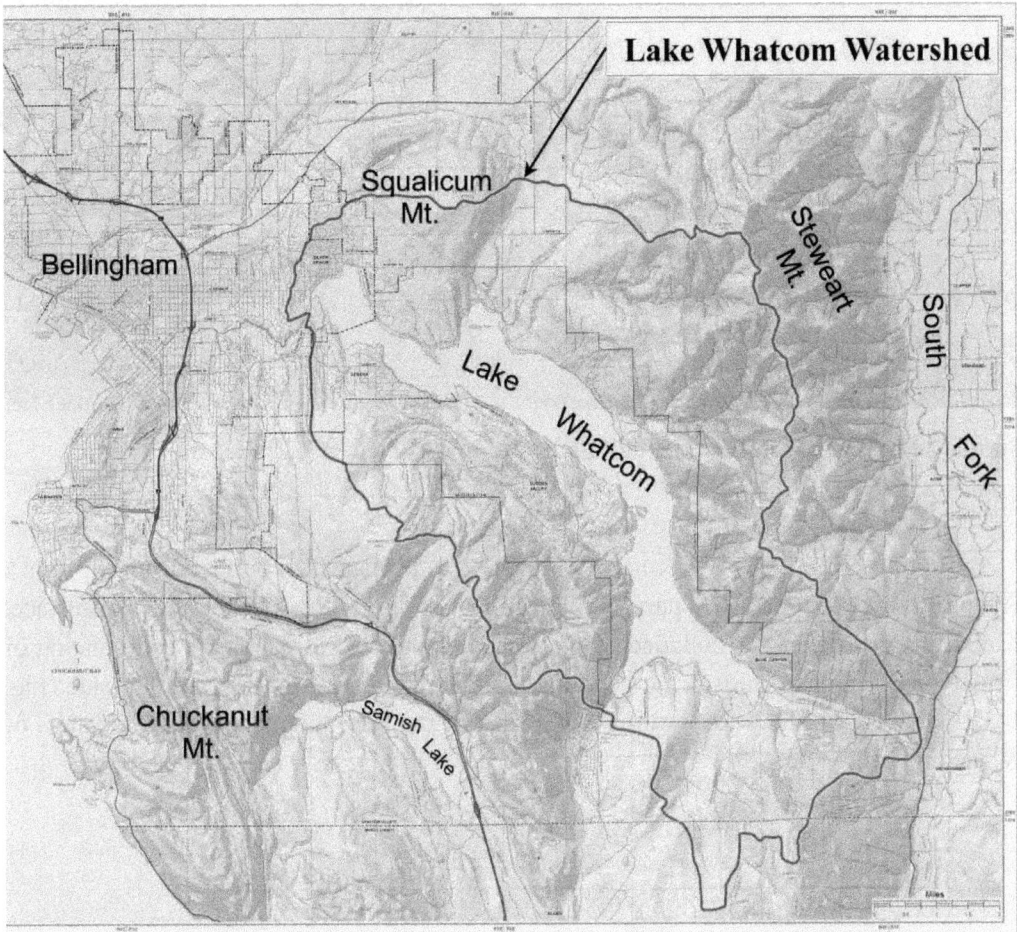

Figure 241. Lake Whatcom watershed. (Modified from City of Bellingham map)

Figure 242. Lake Whatcom. Bellingham in foreground, Mt. Baker on skyline.

Basin Lake 1 of Lake Whatcom is 95 feet deep and Basin 2 is 69 feet deep (Fig. 243). The basins are separated by a narrow rock sill that is only 10 feet below the lake surface. The NE part of Basin 2 is bounded by the Strawberry Point sill that is the continuation of a ridge of Chuckanut sandstone extending into the lake as peninsulas on either side (Fig. 243). The outlet of the lake is at the SW corner of the lake and makes the headwaters of Whatcom Creek.

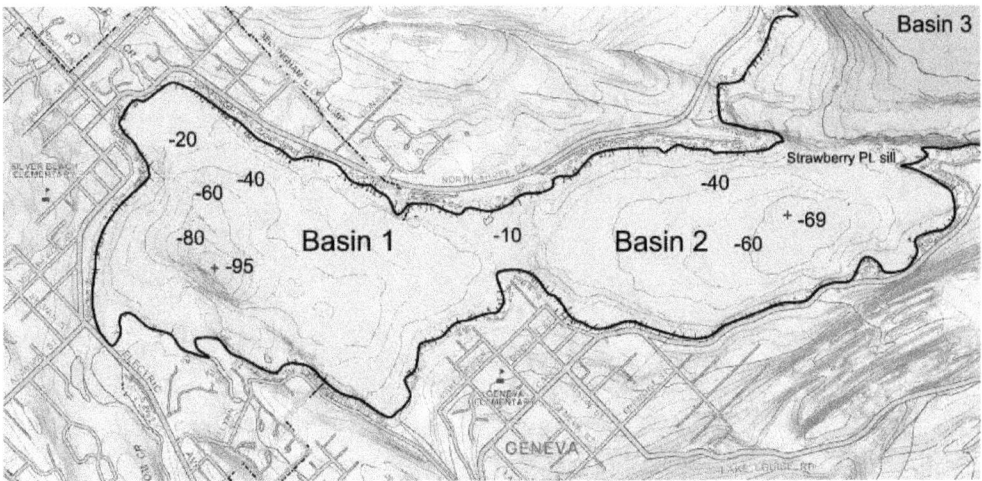

Figure 243. Bottom configuration of Basins 1 and 2 of Lake Whatcom. Numbers are depths below lake level. (Modified from City of Bellingham map)

Basin 3 of the lake is much longer and deeper than Basin 1 and 2. It is bounded on the west end by the Strawberry Point sill and on the east end by another sill 203 feet deep (Fig. 244). At its deepest point, Basin 3 is 266 feet deep.

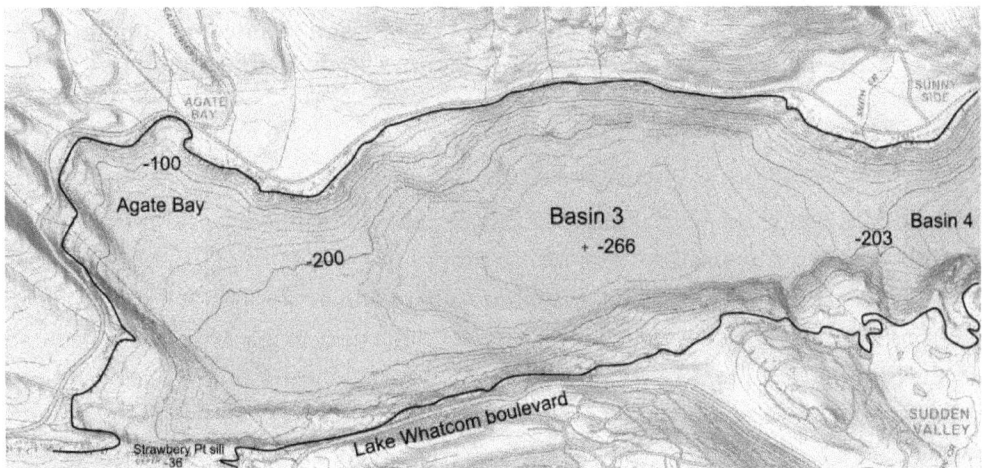

Figure 244. . Bottom configuration of Basin 3 of Lake Whatcom. Numbers are depths below lake level. (Modified from City of Bellingham map)

Basin 4 is the longest and deepest basin of the lake. The deepest part of the lake occurs about midway in Basin 4 (Fig. 245) where it is 328 feet deep, slightly below sea level.

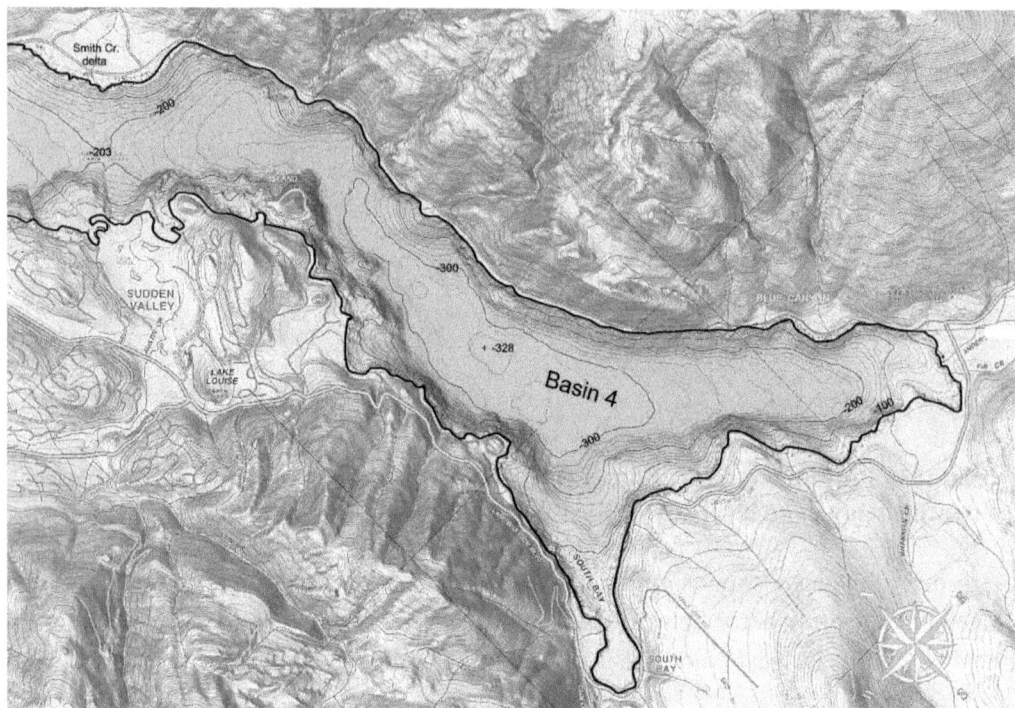

Figure 245. Bottom configuration of Basin 4 of Lake Whatcom. Numbers are depths below lake level. (Modified from City of Bellingham map)

Lake Whatcom has had a varied geologic history. It has been a stream valley, a glacially scoured, trough filled with ice, an arm of the sea, and finally a freshwater lake. Before the first of half a dozen or more Ice Ages spanning the last two million years, a drainage system occupied the Lake Whatcom watershed (Fig. 246). The main trunk stream eroded the original valley floor that was later glaciated and now makes the bottom of the lake, as well as Squalicum valley where the stream exited the foothills into the lowland.

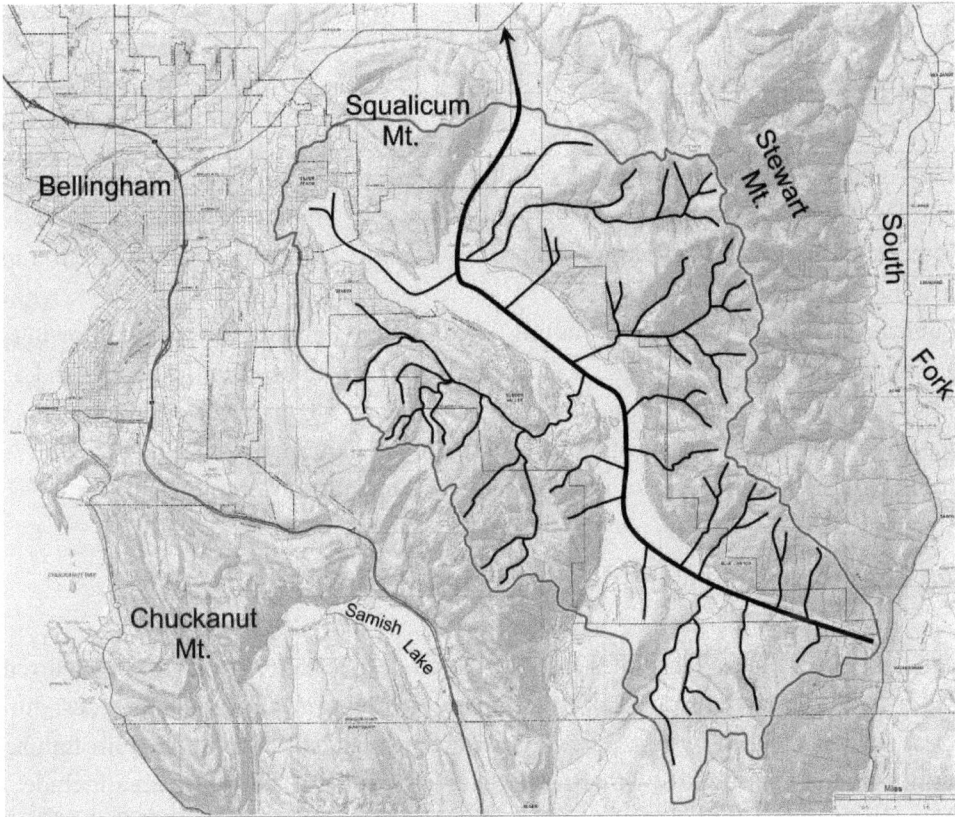

Figure 246. Reconstruction of the pre-glacial drainage that must have occupied the Lake Whatcom watershed.

The second phase in the development of Lake Whatcom was repeated scouring of the valley by 5,000–6,000 feet of ice of the Cordilleran Ice Sheet. At least six major Ice Ages are known in NW Washington in the past two million years. Each time the glacier passed through the area, it deepened and widened the valley floor, making it into a glacial trough that extends below present sea level.

The third phase in the development of Lake Whatcom was invasion of the basin by the sea during the Everson glaciomarine phase about 12,000 years ago. Marine shells in glaciomarine stony clay occur at a few places along Lake Whatcom Boulevard and nearby areas at elevations of 400 feet, about 100 feet above present lake level. At that time, an arm of the sea must have occupied the Lake Whatcom basin.

The fourth phase in the development of Lake Whatcom was emergence of the area from beneath the sea about 11,500 years ago. As relative sea level dropped below about 300 feet, the Lake Whatcom basin was cut off from the sea and the salt water gradually became diluted into fresh water that occupies the basin today.

233

CHAPTER 6

LANDSLIDES

The force of gravity continuously, sometimes catastrophically, wears down mountains and valley sides. Gravity drags everything constantly downward, causing downslope movement of material that varies from slow, subtle, continuous creeping to rapid, devastating landslides.

Landslides consist of rapidly falling, sliding, or tumbling rocks or flowing of unconsolidated debris. Landslides occur abruptly, often with devastating results and are a significant geologic hazard throughout the world. In the United States alone, they are estimated to cause between one and two billion dollars in property damage annually. As a geologic hazard, landslides occur more frequently than volcanic eruptions or earthquakes.

Deep seated bedrock landslides related to earthquakes

At least 13 very large landslides that originated deep within bedrock have occurred in the valley walls along the Nooksack River, many involving whole mountain sides, during the past several thousand years. The largest of these landslides resulted from failure of the entire side of Church Mt. Other large landslides in the surrounding area include: (1) the Van Zandt slide, (2) the Racehorse Creek slide, (3) the Slide Mt. slide, (4) the Canyon Creek slide complex, and (5) five other landslides less than one square mile (Fig. 247).

The occurrence of these large bedrock landslides in an area of recent seismic activity appear to indicate that the slides were triggered by local, shallow, earthquakes in the foothills between the towns of Deming and Glacier.

Historic seismic records show that 245 earthquakes occurred within the area of the huge landslides in the past 40 years, compared to only 94 in the area immediately to the southwest where no large, deep–seated, bedrock landslides have occurred although these areas have the same rock type (Chuckanut sandstone), geologic structure, slope steepness, vegetation, and climate. Earthquakes in the Deming region have released about 10 times more energy than earthquakes in the no–landslide area.

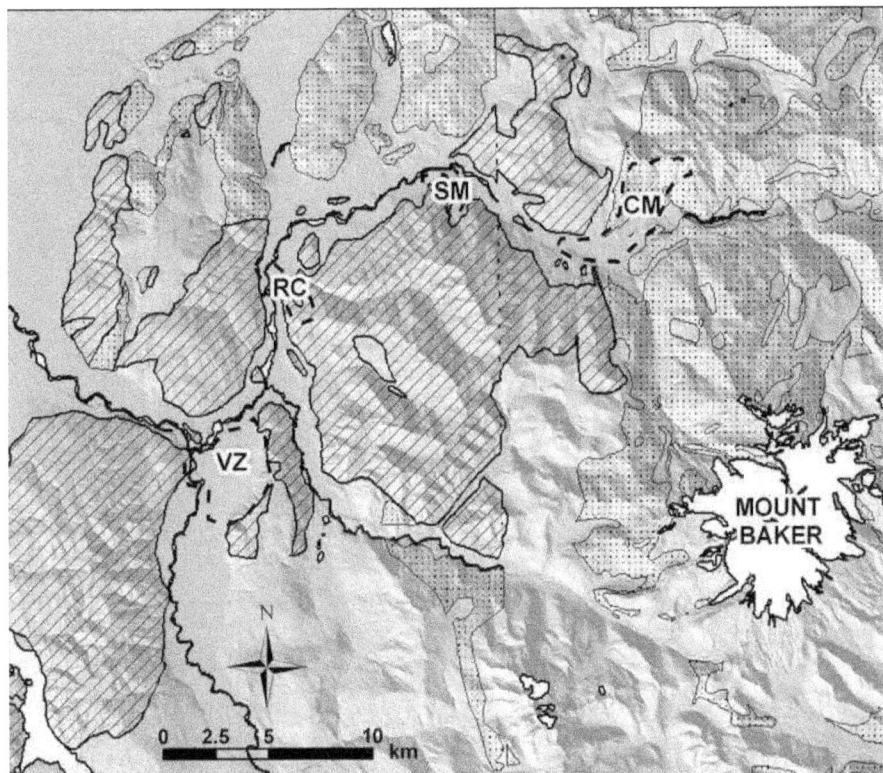

Figure 247. Map showing mega-landslides in the Nooksack drainage and outcrop area of Chuckanut sandstone (diagonal line pattern). VZ–Van Zandt landslide; RC–Racehorse Creek landslide; SM–Slide Mt. landslide; CM–Church Mt. landslide. (From Easterbrook et al., 2007)

Church Mt. landslide

The largest of the landslides came from failure of the entire side of Church Mt. near the town of Glacier (Fig.248). It filled the North Fork of the Nooksack drainage for a distance of seven miles, was up to 1.6 miles wide and up to 312 feet thick.

When driving up the Mt. Baker highway above Maple Falls, the landslide is first encountered at the bridge crossing the Nooksack River below the town of Glacier. From there, past Glacier to the bridge across the river at Douglas fir campground, the topography consists of irregular mounds on the surface of the landslide (Fig. 249)

Figure 248. The Church Mt. mega-landslide above the town of Glacier.

Figure 249. Mound of landslide debris exposed in a roadcut along the Mt. Baker Highway below the town of Glacier.

The Nooksack River was temporarily blocked by the slide but has since re-established its course by eroding through the landslide. A rather extraordinary feature of the Church Mt. landslide is that it consists of three distinct compositional zones of the slide that correspond to the sequence of rock layers that make up Church Mt. The lower half of the landslide consists entirely of rubble from Nooksack Group sandstone, the middle part consists entirely of Chilliwack sandstone and shale, and the upper portion consists of Chilliwack volcanic rocks (Figs. 250, 251). Apparently the entire side of Church Mt. slid down into the Nooksack valley as a large coherent mass so that the rocks maintained their relative stacking order as they slid off the mountain into the valley.

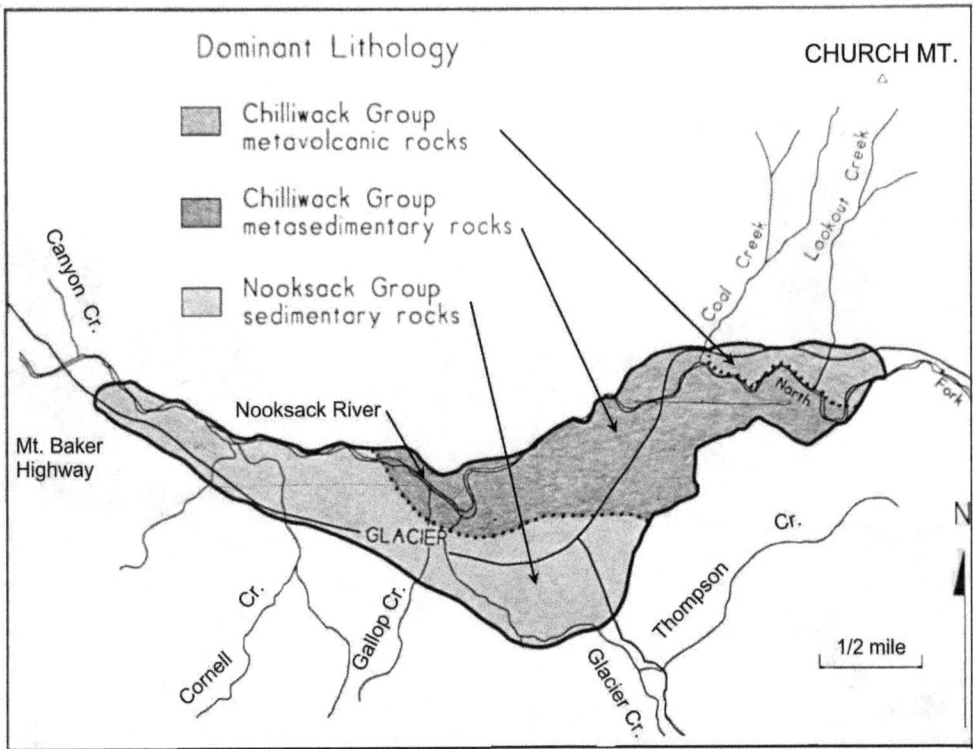

Figure 250. Map of the Church Mt. landslide. (Modified from Carpenter, 1993)

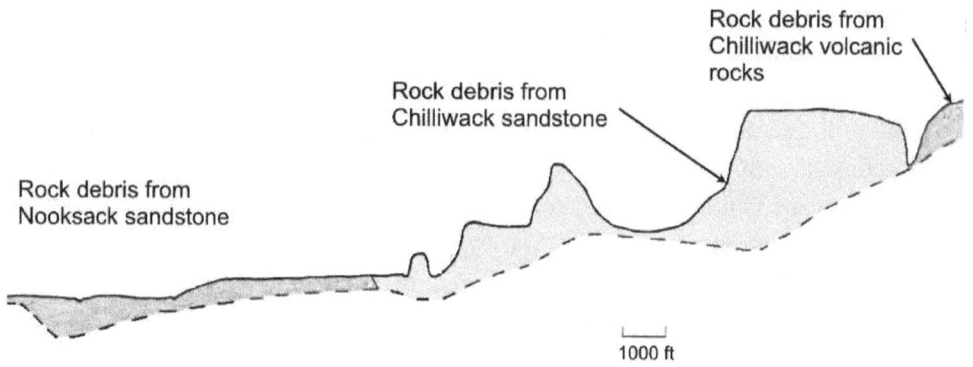

Rock debris from
Chilliwack volcanic
rocks

Rock debris from
Chilliwack sandstone

Rock debris from
Nooksack sandstone

1000 ft

Figure 251. Geologic cross section through the Church Mt. landslide showing the distinct compositional zones of the slide. (Modified from Carpenter, 1993)

Landslide debris

2500 year old log

Figure 252. Large log buried in the Church Mt. landslide near Glacier. The age of the log is 2,500 radiocarbon years.

238

Van Zandt landslide

The Van Zandt landslide is a 2.5 mile–long, 1,800–foot–high failure of the western side of a ridge known as the Van Zandt Dike near Deming (Fig. 253, 254). The slide traveled about two miles into the Nooksack Valley where many huge boulders of Chuckanut sandstone cover an area of about four square miles. The slide scar is so wide that it makes up most of the ridge. The valley floor below the slide is covered with large boulders of Chuckanut sandstone that extend all the way across the valley to the west side where the Nooksack River has cut through the toe of the landslide. A radiocarbon date of 1,600 years was obtained from a core in a bog on the landslide surface, but this is only a minimum age and the slide could be older.

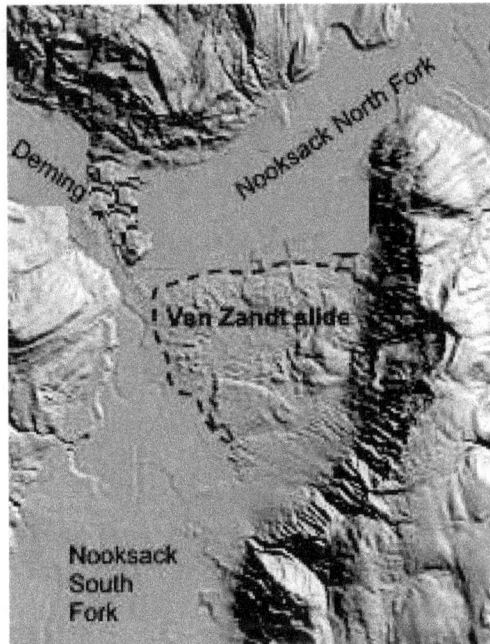

**Figure 253. The Van Zandt landslide near Deming.
The slide is about two miles across. (Modified from
USGS lidar image provided by Whatcom County)**

239

Figure 254. Van Zandt landslide east of Deming.

Racehorse Creek landslide

The Racehorse Creek landslide is about two miles long and one mile wide (Fig. 255). It covers the valley floor and must have temporarily dammed the Nooksack North Fork. The distal portions of the slide are mantled with many huge boulders of Chuckanut sandstone. A deep slide scar occurs on the east side of the valley, extending to the crest of the ridge.

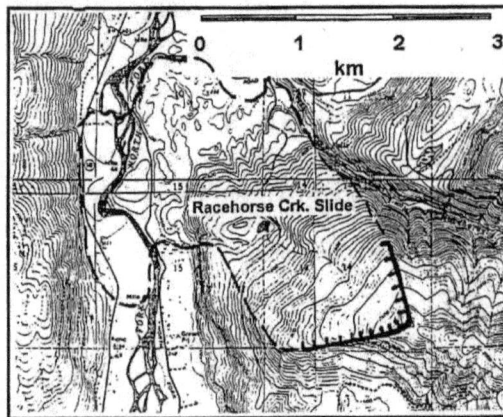

Figure 255. Racehorse Creek landslide south of Kendall.

Slide Mt. Landslide

The Slide Mt. landslide is approximately 2 1/2 miles upstream from the town of Maple Falls in the Nooksack North Fork valley (Fig. 256). It originated in Chuckanut sandstone on Slide Mt. and is about 1.5 miles wide and three miles long.

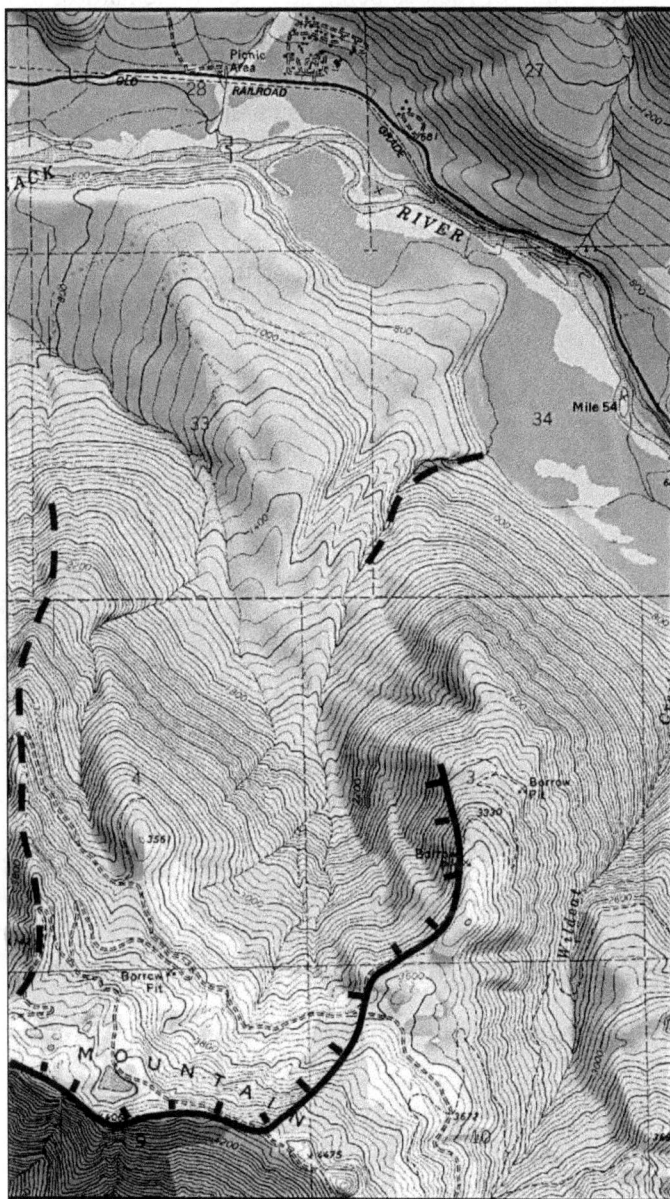

Figure 256. Slide Mt. landslide.

Shallow Landslides

Rockslides

Although we often think of rock as being solid, even the hardest of rocks can be split by fractures and can slide along them. Rockslides occur where blocks of rock break away and slide down the surface of a fracture plane or along bedding (Fig. 257). Rock slides along joint and bedding planes in the Chuckanut sandstone are a perpetual problem in road cuts along I-5 between Bellingham and Lake Samish and along Chuckanut Drive.

Fig. 257. Rock slide on fracture planes in Chuckanut sandstone along I-5 south of Bellingham.

A good example of a larger rockslide occurs above Bagley Lakes at Heather Meadows at the end of the Mt. Baker highway. A large scar is apparent on the side of Mt. Herman (Fig. 258) where a large mass of rocks has broken away and slide into the valley holding Bagley Lakes. The rock slide here is in fractured Chilliwack volcanic rocks.

Figure 258. Scar on Mt. Herman from large rock slide. View from Heather Meadows.

During a heavy rainstorm in January, 2009, a large mass of Chuckanut sandstone in the upper reaches of Racehorse Creek broke loose along a coal bed and slid downslope into Racehorse Creek (Figs. 259, 260). The initial failure began as a moving slab of sandstone about 80 feet thick sliding on the surface of the coal bed, then broke up as the mass slid into Racehorse Creek. It is a good example of a bedding plane failure.

Figure 259. Large rock slide along a bedding plane near Racehorse Creek north of Deming. A huge block of Chuckanut sandstone broke loose and slid along a coal bed into the valley below. (Photo by Jack Powell)

Figure 260. Scarp left by the large rock slide shown in Figure 259. Note the people at the top of bluff for scale. The scarp is about 80 feet high. (Photo by Jack Powell)

Slump

Slump is the downward sliding of a mass of slope material along a curved, concave–upward plane of failure, usually with backward rotation of the original land surface at the top of the failure. A bluff above the moving slide material marks the upper part of the plane of failure after a slump (Fig. 261). The toe of slump blocks may evolve into earthflows or mudflows if the materal is saturated (Fig. 262).

Figure 261. Slump in hillslope along the valley side of the Nooksack Middle Fork.

Slumping often occurs in unconsolidated clay, which has lower internal strength when saturated with water, making them plastic. Slumping is especially common during heavy rains because of this. Much of Whatcom County is underlain by Bellingham glaciomarine stony clay that is prone to slumping when saturated (Figs. 262–264).

Figure 262. Slump of glaciomarine stony clay at a sea cliff near Birch Point north of Birch Bay. Note the flowage of material at the toe of the slump.

Figure 263. Slumping of Bellingham stony clay has undercut a building along the Nooksack River at the end of the Smith Road.

Figure 264. Slumping of clay banks along the Nooksack River at the end of the Smith Road. The river undercut the clay on the outside of a bend, resulting in a large slump that changed the course of the river.

248

Debris Torrents

Debris torrents are large masses of water, logs, and rock debris that hurtle down stream channels at high velocity with very destructive effects (Figs. 265-270), driven by water and momentum. They differ from mudflows in that mudflows are driven by viscous flowage of clayey material whereas debris torrents typically consist mostly of logs and water driven by gravity and momentum.

Fig. 265. House crushed by 1983 debris torrent in Smith Creek.

An eyewitness account of a debris torrent in Smith Creek on the north side of Lake Whatcom described a debris torrent in 1983 as an "explosion of logs and debris" tossed 25 feet in the air as the debris crashed into a bridge. Then it came down, and two seconds later, this mass of debris was rolling across the road. Looked like 50 bulldozers just pushing a big bunch of spaghetti (logs), just going end over end."

Fig. 266. House destroyed by 1983 debris torrent in Smith Creek.

Fig. 267. House destroyed by 1983 debris torrent in Smith Creek.

Fig. 268. House demolished by 1983 debris torrent in Smith Creek.

Figure 269. House floating in Lake Whatcom at Blue Canyon, pushed into the lake by a 1983 debris torrent

Fig. 270. House floating in Lake Whatcom after being pushed in the lake by the 1983 debris torrent in Smith Creek.

252

Figure 271. Huge boulders moved by the 1983 debris torrent in Olsen Creek on the north side of Lake Whatcom.

Huge boulders were moved downstream during the l983 debris torrent in Olsen Creek (Fig. 271). The boulders in Figure 271 came to rest on green branches, showing they had moved during the debris torrent.

Observations made soon after debris torrents occurred in drainage basins in Whatcom County have revealed that debris torrents typically consist mostly of logs and other vegetative debris left in the channels from previous timber harvesting operations (Fig. 272). The violence of debris torrents scours stream channels down to bedrock, tearing up trees, roots, and soil, leaving bare rock chutes scoured 15 to 30 feet from the valley floor up the valley sides (Fig. 273). The origin of debris torrents is always traceable to a slope failure above a channel in the upper part of a stream drainage.

Precipitation records indicate that storms capable of producing debris torrents occur every 2 to 15 years when more than four inches of rain falls in 24 hours where past logging operations have left large amounts of logging woody debris on steep slopes. The woody debris eventually finds its way into stream channels (Fig. 272) where it awaits rapid movement downstream by debris torrents. Once a debris torrent is generated, logs are readily available for incorporation in the debris torrent (Fig. 274). In studies of debris torrents in 12 drainages in western Washington, logging debris was responsible for virtually all of the damage associated with the torrents. The influence of human activity

253

in causing debris torrents is well shown in that most of the logs carried by debris torrents have sawed ends from logging. Counts of logs with sawed ends in debris torrents show that as much as 90% of the logs were from logging.

Fig. 272. Stream channel choked with logging slash.

Figure 273. Channel of Smith Creek after the 1983 debris torrent. Before the debris torrent, the channel was full of logs from previous timber harvesting. The debris torrent scoured out all of the logs and trees, brush, and soil down to bedrock.

Figure 274. Logging slash carried by the 1983 debris torrent in Olsen Creek on the north side of Lake Whatcom.

The points of origin of debris torrents are slope failures at logging roads in the upper part of tributary systems (Fig. 275). Sliding failure of road-fill is well demonstrated where sections of road have collapsed into stream gullies, initiating debris torrents downstream. Proof that road-fill failures cause the debris torrents is shown by the fact that stream channels below road-fill failures are scoured while those above the failures are not, and no debris torrents occur without slope failures. Figures 276 to 278 show ample proof of the origin of debris torrents in Sygitowicz Creek during 1983 debris torrents. All of the debris chutes that fed the debris torrent in the creek originated at logging road slides.

The cause of debris torrents is (1) accumulation of logs in stream channels as a result of logging activities, (2) rainfall of about 4^+ inches in 24 hours (or rain plus melting snow), and (3) slope failures, almost entirely from logging road failures.

Figure 275. Incipient failure of a logging road.

Figure 276. Debris chutes from logging road slides that provided logging slash for the destructive 1983 debris torrents in Sigitowicz Creek.

Figure 277. Debris chute from a logging road slide that provided woody material for the destructive 1983 debris torrent in Sigitowicz Creek

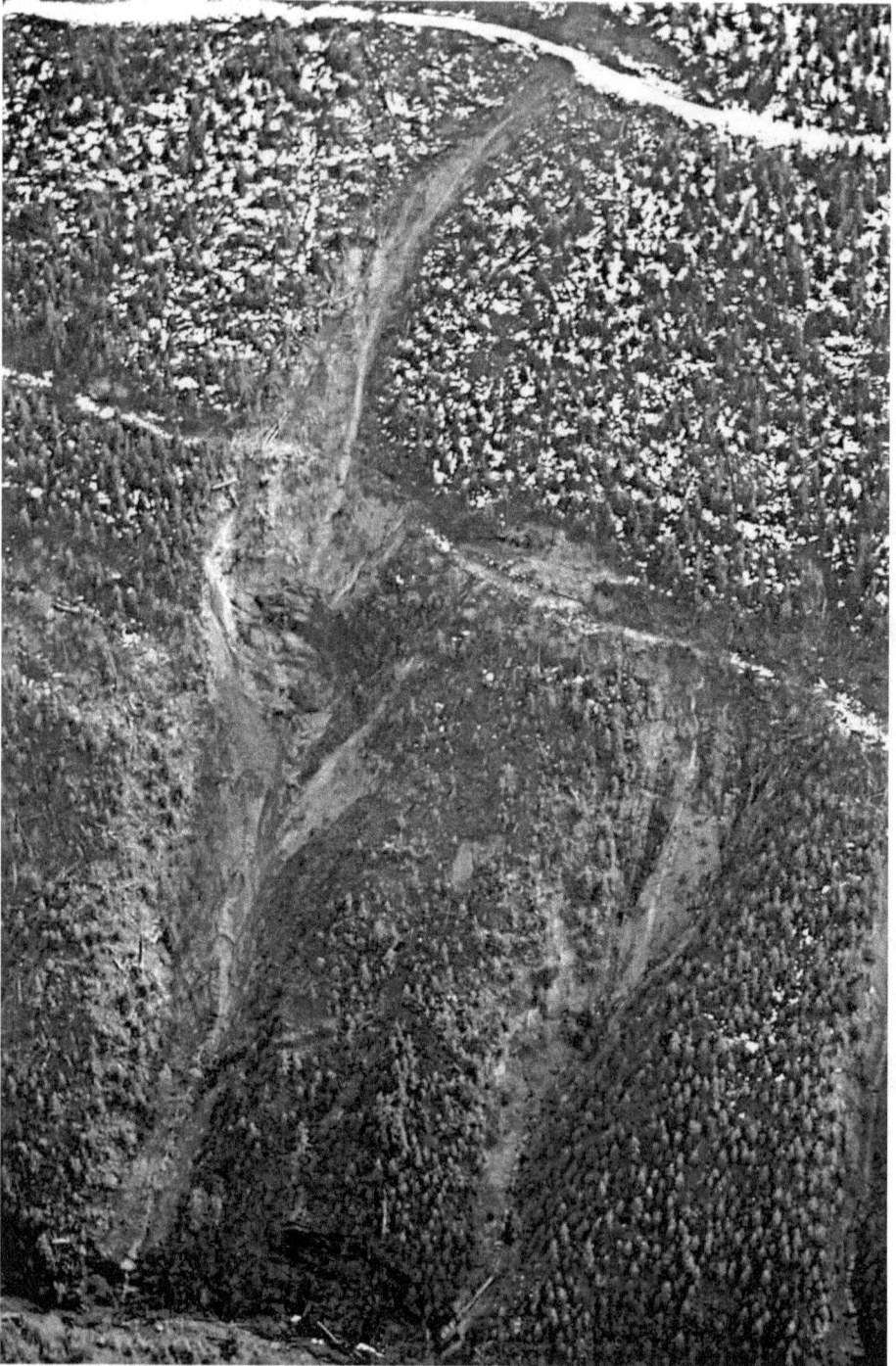

Figure 278. Debris chutes from logging road slides that provided woody material for the destructive 1983 debris torrents in Sigitowicz Creek

CHAPTER 7

GEOLOGIC HISTORY OF THE NOOKSACK RIVER—A TALE OF REMARKABLE CHANGES

Watching the Nooksack River over a lifetime gives one the impression that the river has been much the same throughout its history, but a closer look at the geologic changes that have occurred over just the past 20,000 years reveals remarkable changes. At one time or another within the past 20,000 years, the Nooksack River has flowed into Lummi Bay, into the Fraser, River, into the Skagit River, been an arm of the sea, been covered by more than a mile of glacial ice, had a substantially higher discharge of water, and lost about half of its watershed.

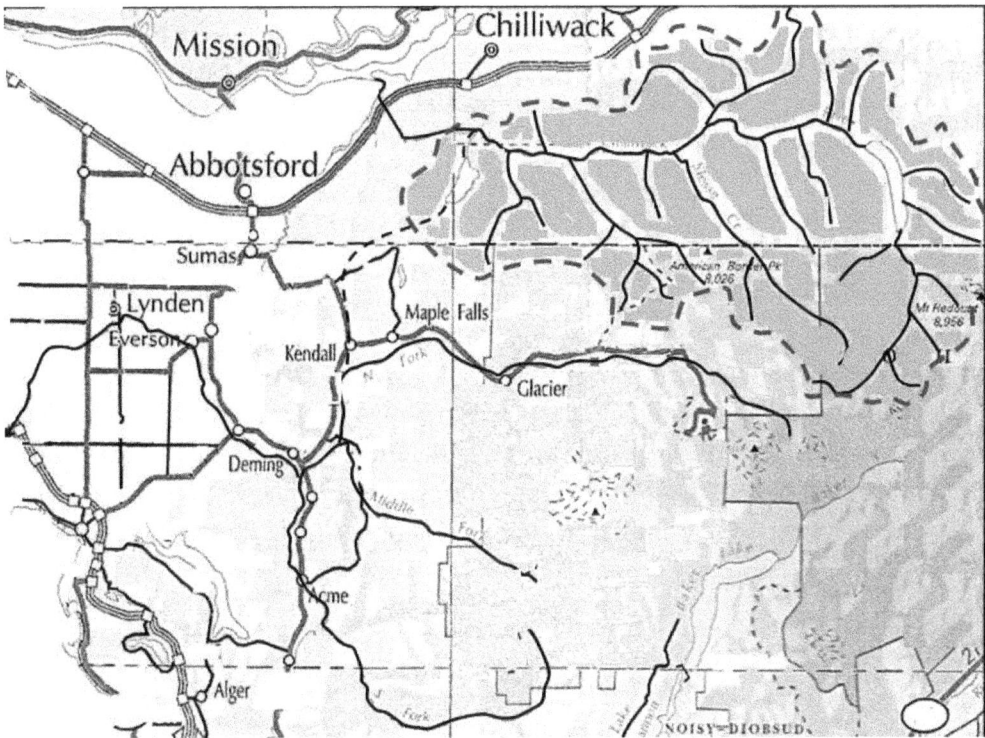

Figure 279. Map of the ancestral Nooksack River drainage. The dark area in the northeastern part of the map is now the Chilliwack River drainage system that flows into the Fraser River, but it was once part of the Nooksack watershed through Columbia Valley prior to its diversion about 12,000 years ago.

259

NOOKSACK RIVER

The ancestral Nooksack River prior to 20,000 years ago was much larger than the present river. The Nooksack drainage formerly included the large watershed of what is now the Chilliwack River (Fig. 279) that flowed from Canada down Columbia valley to join the rest of the Nooksack drainage at Kendall. The area of the Chilliwack drainage basin is approximately the same as the area of the present Nooksack drainage, so prior to 20,000 years ago, the Nooksack watershed was about twice what it is now. Flood discharges of the Chilliwack River are similar to those of the Nooksack River, so at that time, the Nooksack was probably about twice as big as it is now.

The Cordilleran Ice Sheet advanced from Canada across Whatcom County on its way to the southern Puget Lowland on at least six occasions during the Ice Ages. Each time, the drainage of the Nooksack River was covered by more than a mile of glacial ice and the river totally obliterated. Upon melting away of the ice sheet at the end of each Ice Age, the Nooksack River reoccupied its valley.

During deposition of the Bellingham stony clay 11,700 radiocarbon years ago near the end of the last Ice Age, relative sea level was about 500-600 feet above present sea level and an arm of the sea filled the Nooksack Valley from the lowland upvalley at least as far as Welcome and probably higher.

During the subsequent emergence of the lowland from the sea about 11,500 radiocarbon years ago, a valley glacier occupied the Chilliwack valley, terminating at Cultus Lake just north of the Canadian border. At that time, a remnant of the Cordilleran Ice Sheet in the lowland blocked the Nooksack valley west of Deming, forcing meltwater from Columbia Valley southward through what is now the Nooksack South Fork and into the upper Samish River valley. As shown by the unusually large bends in the Samish valley, this stream had a significantly higher discharge than the Samish River. The stream exited the Samish River valley just north of Burlington and flowed into the Skagit River (Fig. 280).

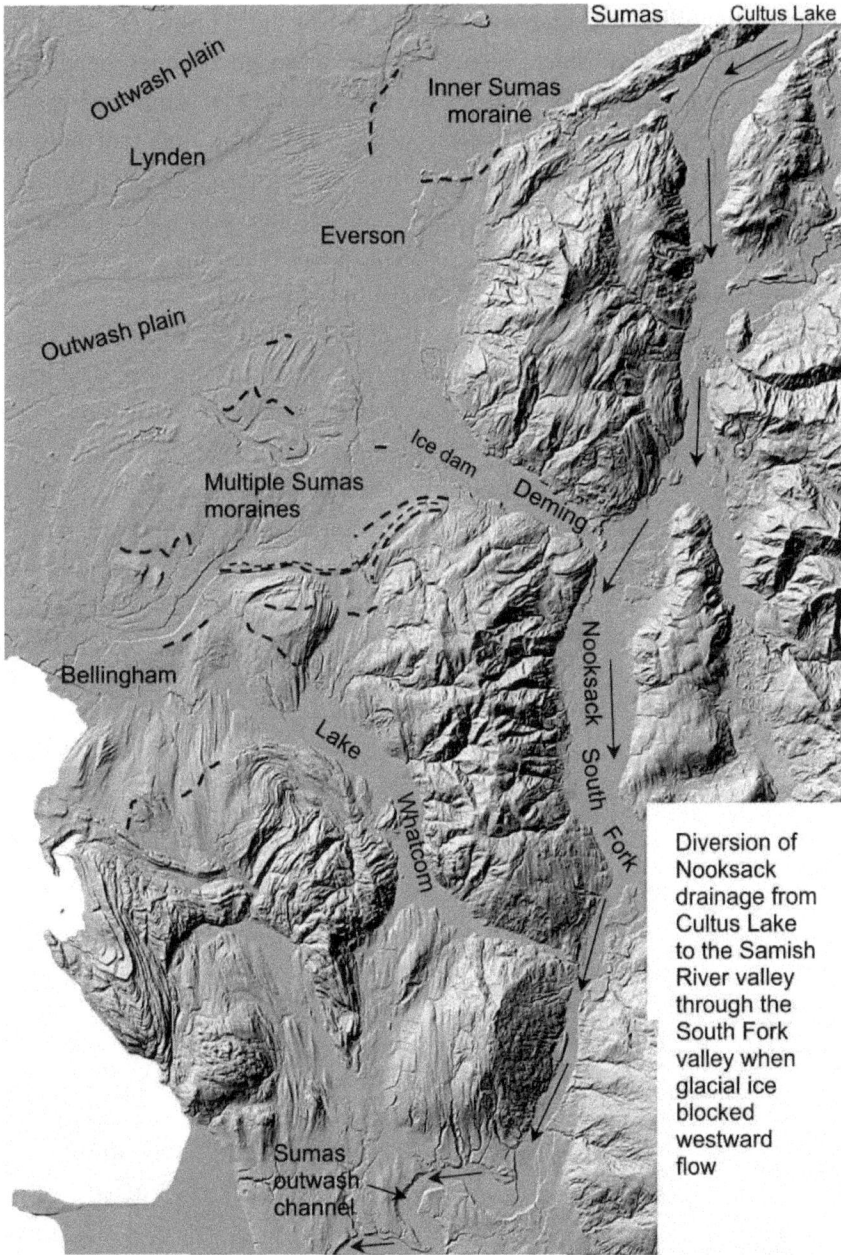

Figure 280. Former course of the meltwater stream from a glacier at Cultus Lake in British Columbia down Columbia Valley to Kendall, southward in the Nooksack South Fork valley, and then down the Samish River valley to a delta 100 feet above sea level just north of Burlington. (Modified from USGS lidar provided by Whatcom County)

261

When ice in the Whatcom County lowland and in the Chilliwack valley at Cultus Lake (Figs. 224-226) melted back, two important drainage changes occurred; (1) the glacier at Cultus Lake pulled back from the embankment of sand and gravel it had previously deposited at the terminus of the glacier, leaving a 400 foot high bluff that caused the Chilliwack drainage to be diverted into the Fraser River north of Sumas, thus robbing the Nooksack of half of its former drainage area, and (2) ice in the lowland melted back from the Nooksack valley near Deming, allowing the Nooksack drainage to resume flowing westward across the Whatcom County lowland.

As shown by the presence of pebbles of Mt. Baker lava in the subsurface sediments of the Sumas valley, the ancestral Nooksack at times flowed northward past Sumas into the Fraser River. This probably happened repeatedly. The Sumas valley is now occupied by the Sumas River, which begins its course on the Nooksack River floodplain several miles south of Everson, flows on the Nooksack floodplain, then branches to the northeast at Everson and flows in the Sumas valley into the Fraser River. During times of flooding of the Nooksack River, flood waters sometime spill into the Sumas Valley and flood the town of Sumas. During the early part of the 20th century, the northern part of Sumas valley was occupied by Sumas Lake, a shallow lake with a good deal of surface area (Fig. 281).

Figure 281. Sumas Lake near the town of Sumas in the early part of the 20th century.

NOOKSACK DELTA

In early post-glacial time, the Nooksack flowed into the sea probably near Ferndale where it discharged its load of silt and sand into the sea and began to build a delta. A 5,000-year-old archeological site near Ferndale contains marine clam shells suggesting that the shoreline was not far distant from there at that time. As the Nooksack continued to dump its sediment load into the sea and build its delta seaward, it gradually extended the land seaward until it reached what is now the Lummi Peninsula (Fig.282). Until then, the Lummi Peninsula had been an island, but as the delta expanded, it annexed the island to the mainland. Thereafter, the Nooksack alternated between Lummi Bay and Bellingham Bay. When the distance from the junction to the mouth of one of the courses became substantially longer than the other, flow was diverted to the shorter, steeper route. By alternately occupying both courses, twin deltas were built, one into Bellingham Bay, the other into Lummi Bay.

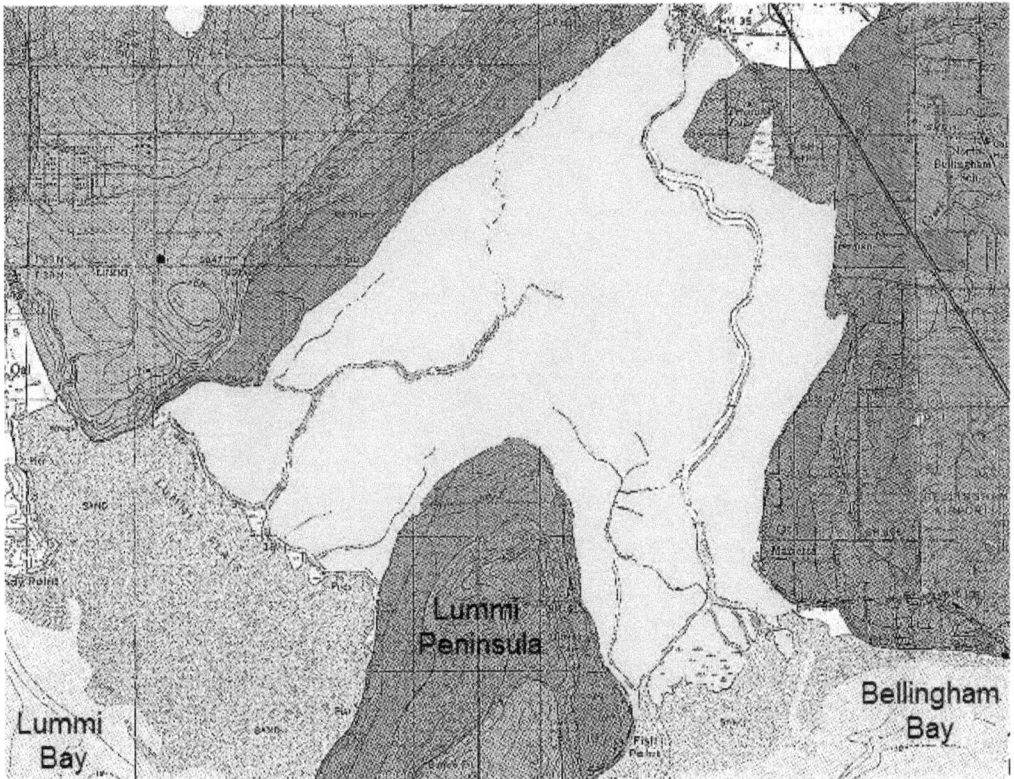

Figure 282. Bellingham Bay and Lummi Bay deltas of the Nooksack River. (Modified from USGS map)

As the Nooksack delta built seaward, the part of the delta above high tide is easily visible and that is what shows up on a map. However, the front of the advancing delta is actually below sea level, extending some distance offshore (Figs. 283-286). The significance of this is that the delta is rapidly extending into Bellingham Bay where it affects the port facilities.

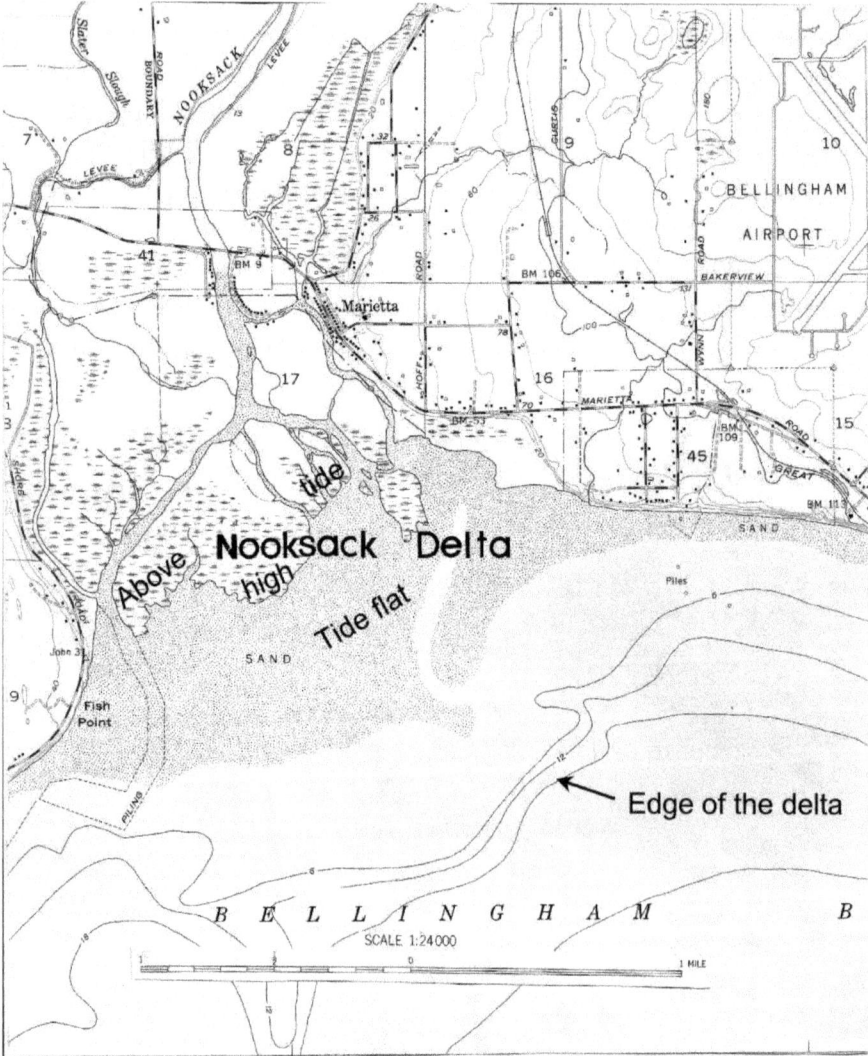

Figure 283. Map of the Nooksack delta. Note that the delta extends more than a mile beyond the visible part of the delta above high tide. (Modified from USGS map)

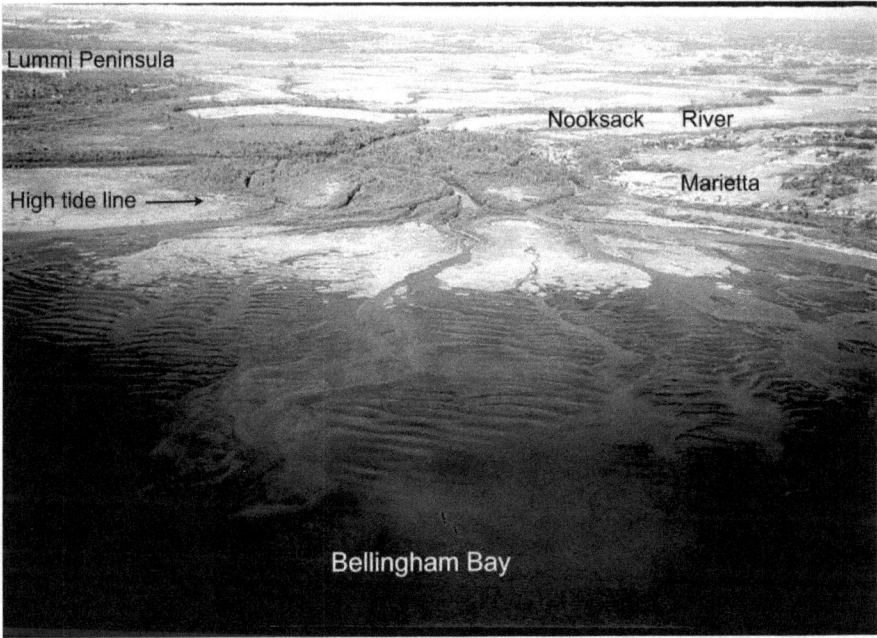

Figure 284. Nooksack delta building into Bellingham Bay.

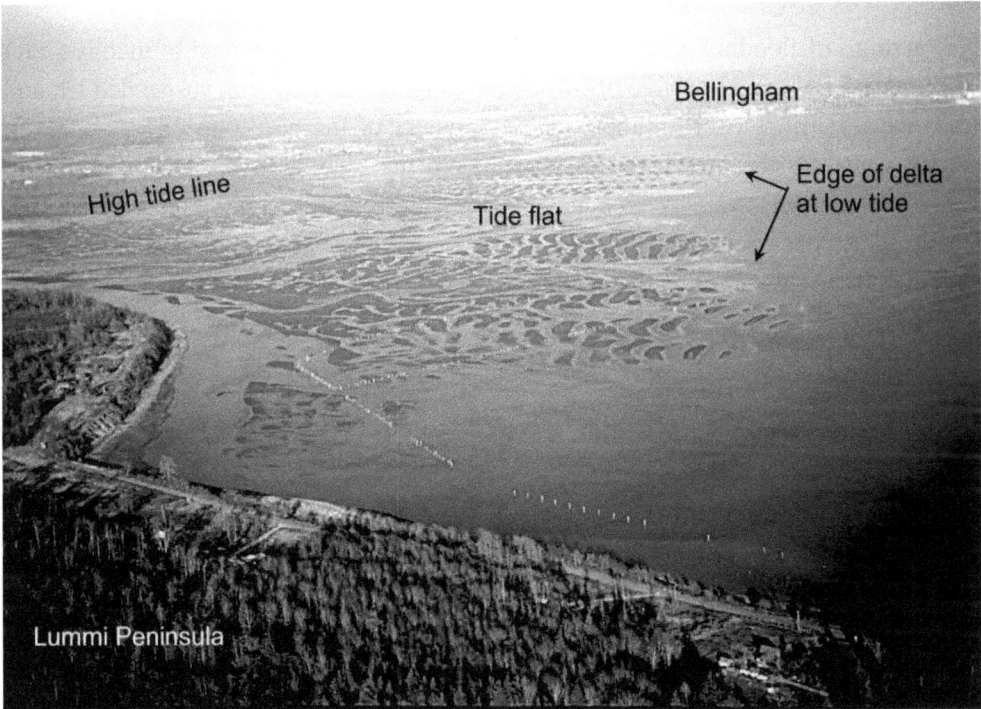

Figure 285. Portion of the Nooksack delta below high tide.

265

Figure 286. Nooksack delta at low tide showing the broad tide flat that extends seaward about one mile from the high tide line.

During times of flooding, the Nooksack River discharges a great deal of muddy water into Bellingham Bay ,and a wedge of fresh water, highly charged with suspended silt and clay, advances seaward over the more dense saltwater of the bay (Fig. 287). The sharp line which separates the brown muddy water from the blue saltwater moves gradually seaward, carrying silt and clay well beyond the limits of the delta. The silt and clay eventually settles out on the floor of the bay and is distributed along the margins by wave and current action.

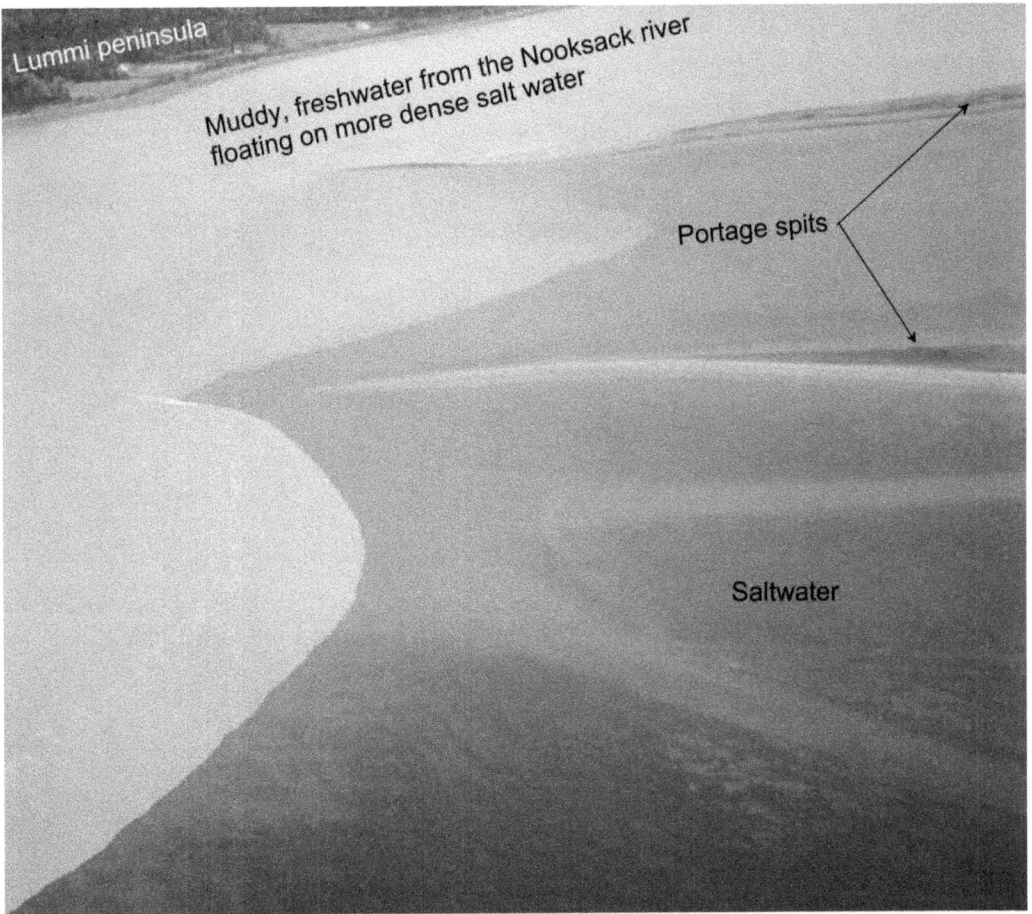

Figure 287. Muddy freshwater from the Nooksack River floating on more dense saltwater in Bellingham Bay between the Lummi Peninsula and Pt. Francis. (Photo by D.A. Rahm)

Figure 288 shows the rate of extension of the Nooksack delta into Bellingham Bay since 1898. The part of the delta above high tide and above low tide for 1898 and 1969 shows that the portion of the delta above high tide advanced nearly one mile into Bellingham Bay and front of the delta offshore advanced significantly over the 71 year period. The lateral parts of the delta along the sides extend as "wings," which may be seen reaching toward the port facilities of Bellingham. At the turn of the century, there was virtually no tide flat just offshore from the sea cliffs near Bellingham.

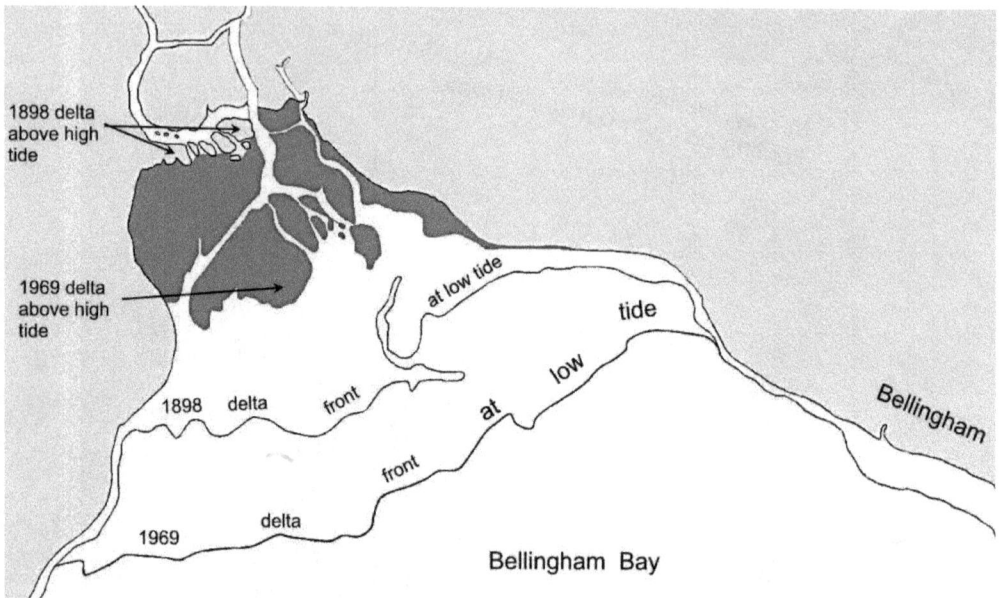

Figure 288. Comparison of the Nooksack delta in 1898 and 1969.

THE NOOKSACK FLOODPLAIN

The Nooksack River floodplain (Fig. 289) has been created by the incessant movement of meanders (bends) back and forth across its valley floor. The meandering of streams has significance impacts on land use because virtually all parts of the floodplain are eventually occupied by the river, bank erosion may be rapid, and channels are deepened considerably during periods of flooding. Encroachment of meanders on man-made features (Fig. 290) may result in substantial damage.

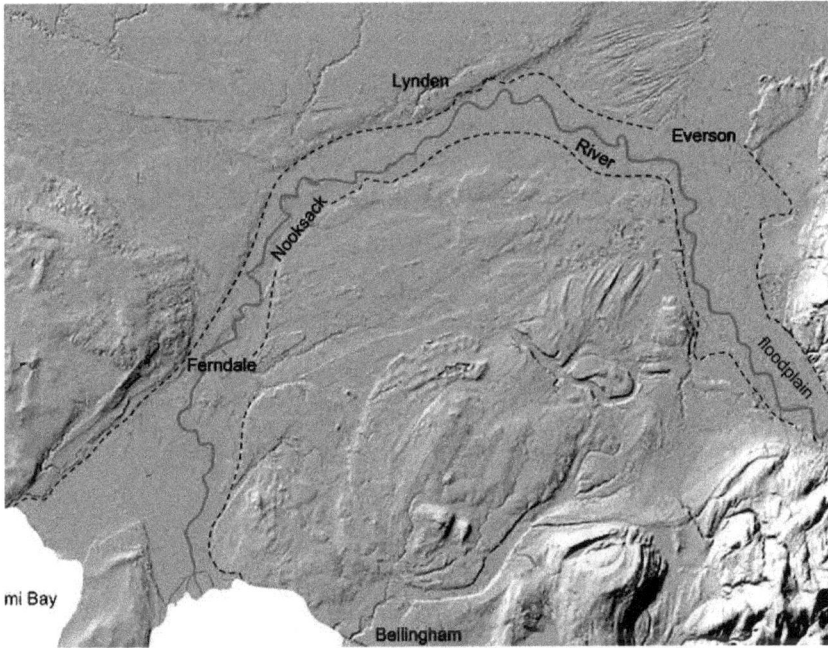

Figure 289. Floodplain of the Nooksack River (dashed lines). (Modified from USGS lidar image provided by Whatcom County).

Figure 290. Undesirable result of meandering of the Nooksack River. Accelerated erosion on the outside of the river bend has undercut the house.

269

The Nooksack River has two channel patterns, meandering and braiding (Figs. 291-292). Meandering channels are characterized by sinuous bends, which move back and forth across the floodplain, whereas braiding channels consist of multiple channels that diverge and reunite repeatedly. Figures 291 and 292 illustrate the two channel patterns. The reason for the two channel patterns is that braided streams form where bank-forming material has little clay or silt and thus has low cohesion and is easily eroded. Meandering streams have bank-forming material that contains cohesive clay that allows a single stable channel to form.

Figure 291. Meandering pattern of the Nooksack River.

Because of bank erosion and deposition of river bars, natural rivers and streams are never straight for any substantial length of their channel. When water flows around a bend in a stream, the water sloshes up on the outside of the channel with a stronger current than on the inside of the bend because of the momentum of the moving water. Thus, the channel is deepened on the outside of the bend and bank erosion becomes concentrated there. Continued bank erosion on the outside of a bend causes the bend to enlarge and to migrate downstream. Material eroded from the outside bank of a bend may be deposited on the inside bank of a meander (bend) farther downstream. Although driftwood and log jams may locally affect stream flow, they are not a fundamental cause of river meanders.

Parts of the Nooksack River have a braided pattern with multiple channels that diverge and reunite repeatedly (Fig. 292). These parts of the channel migrate back and forth more readily than meandering channels.

Figure 292. Braided pattern of the Nooksack River, characterized by numerous channels that branch and reunite.

Like all rivers, the Nooksack the amount of water flowing in the river depends on precipitation and groundwater. The river level is higher during the wet season (November to June) and lower during the dry season (summer). Occasionally, rainfall from large storms will cause the river to top its banks and spill onto the floodplain (Fig. 293). Erosion is heightened by flooding, often causing damage to buildings and bridges (Fig. 294).

Figure 293. 1975 flooding of the Nooksack River between Ferndale and Bellingham and Lummi Bays.

Figure 294. Collapse of the bridge at Nugents Corner during flooding.

CHAPTER 8

BEACHES AND SPITS

The incessant action of waves breaking against the shorelines along the shores of Georgia Strait and the adjacent waterways has produced many miles of beaches and several prominent spits in Whatcom County. Wave erosion undercuts the base of slopes along the shore and sea cliffs are formed which retreat landward under continued wave attack. Sand and other particles are dragged back and forth, pushed up the beach with each wave and pulled seaward again as the water runs back. The net effect is analogous to a great saw, held horizontally, the teeth of which are the loose particles moved by the waves. Prolonged erosion causes retreat of sea cliffs and creates wave-cut benches, beaches, and tide flats parallel to the shoreline

Rock debris derived from erosion of the sea cliffs is transported seaward or laterally along the shore until the motion of the water is no longer able to keep the particles moving. Waves which impinge on the shoreline at an angle push sand grains up the beach at an angle to the shoreline but as the water runs back down the sloping beach they travel directly down the slope. Each wave imparts a lateral component of movement to the particles which march along the shoreline until converging at the head of a bay (Fig. 295) or the deep water of a bay is encountered.

At Birch Bay, sand eroded from Birch Point and Point Whitehorn is carried laterally along the shoreline by wave action until it reaches the head of the bay and accumulates as a broad tidal flat. Beaches such as these are known as bayhead beaches because the sand is trapped at the head of the bay.

At Semiahmoo spit (Figs. 295, 296), sediment is carried laterally along the shoreline by wave action, but instead of moving into a bay, it is carried into deeper water where it comes to rest on the sea floor. As the sand accumulates at the end of the spit, it causes the spit to extend farther across Drayton Harbor and will eventually close it off. Other examples of spits formed by this process are Sandy Point (Fig. 297) and the double spits off Point Francis in Bellingham Bay (Fig. 298).

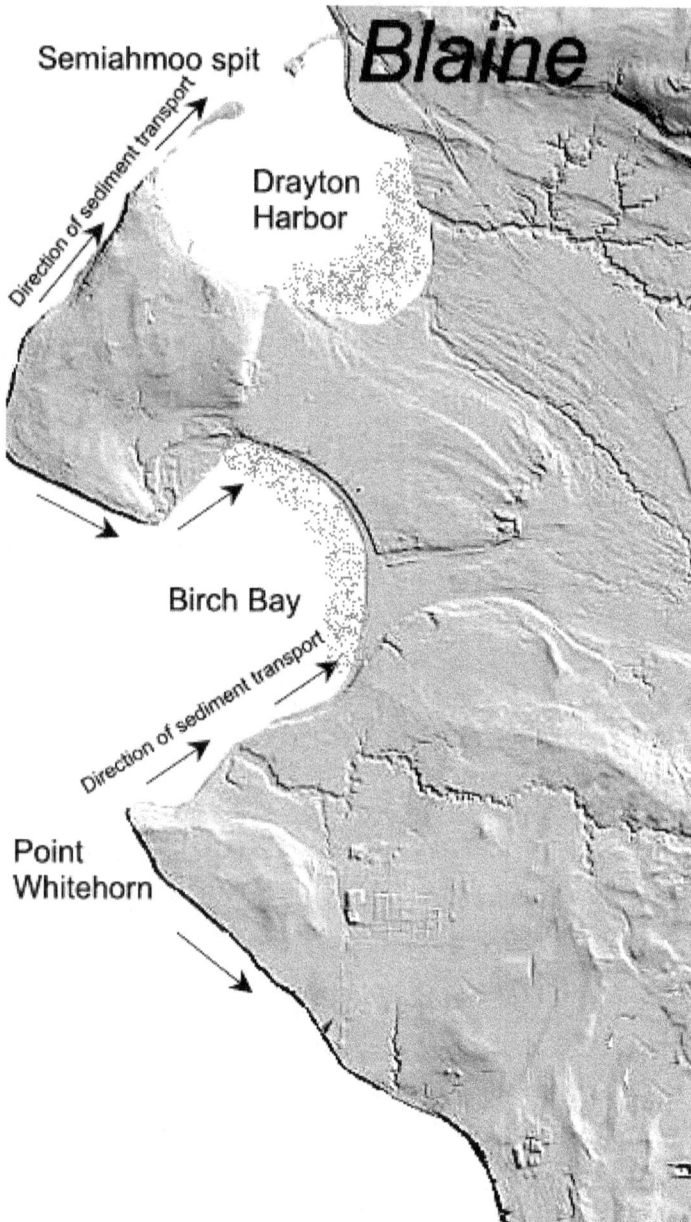

**Figure 295. Lidar image of shorelines near Birch Bay.
(Modified from USGS lidar image provided by Whatcom County)**

Figure 296. Semiahmoo spit extending across Drayton Harbor near Blaine.

Figure 297 Sandy Point spit building into Lummi Bay.

Figure 298. Double spits off Point Francis in Bellingham Bay

GLOSSARY

Ablation The combined processes by which a glacier loses ice.

Ablation area The part of a glacier or snowfield where ablation exceeds accumulation.

Abrasion Wearing away by friction.

Accumulation area The area of a glacier in which annual accumulation exceeds ablation.

Aggradation The process of building up a surface by deposition of sediment.

Alluvial fan Low, cone-shaped deposit formed by a stream issuing from mountains into a lowland.

Alluvium Sand, gravel, and silt deposited by rivers and streams.

Alpine glacier Glaciers occupying mountainous terrain.

Altitude: The vertical distance between a point and mean sea level.

Amphibole A family of silicate minerals forming prismatic or needlelike crystals that contain iron, magnesium, calcium, and aluminum. Hornblende is the most common variety of amphibole. It is dark green to black and forms in many igneous and metamorphic rocks. Actinolite is light green. Blue amphiboles contain sodium and have a blue color.

Andesite Fine-grained, generally dark-colored, volcanic rock. Commonly has visible crystals of plagioclase feldspar. Occurs in lava flows and dikes. The most common rock making up Mt. Baker.

Anticline Structure in which beds dip in opposite directions from the central axis.

Arete A sharp-crested mountain ridge formed by headward erosion of back-to-back cirques.

Ash Volcanic dust particles ejected when a volcano erupts.

Ash flow A turbulent mixture of volcanic fragments, lava, and gas erupted violently from a volcano and flowing down slope, usually at high velocities.

Axis of a fold   The line following the apex of an anticline or the lowest part of a syncline.

Bar   An embankment of sand or gravel deposited on the floor of a stream, in the sea, or in a lake.

Basal till   Poorly sorted mixture of sand, silt, clay, pebbles, cobbles, and boulders deposited from the base of a glacier.

Basalt   Fine-grained, generally black, volcanic rock erupted as lava flows or intruded into fractures to form dikes.

Base level   The level below which a land surface cannot be reduced by running water. Sea level is the ultimate base level for streams.

Basement complex   Very old crystalline igneous and metamorphic rocks that lie beneath the oldest sedimentary rocks in a region.

Batholith   Very large mass of slowly cooled, intrusive molten rock such as granite.

Beach   A deposit of wave washed grains along a shoreline.

Bedding   Sedimentary layers in a rock. Beds are distinguished from each other by grain size and composition.

Bedding plane   The planar surface between layers of sedimentary rock.

Bedrock   Continuous solid rock that underlies an area.

Blueschist   A foliated crystalline metamorphic rock rich in blue amphibole formed by high pressure and low temperature.

Braided stream   A stream flowing in multiple dividing and reuniting channels resembling the strands of a braid.

Breccia   Rock made up of angular fragments of other rocks. **Volcanic breccia** is made of volcanic rock fragments generally blown out of a volcano or eroded from it. **Sedimentary breccia** is made of angular pebbles deposited by surface processes.

Calcite   Mineral made of calcium carbonate (CaCOs). Generally white, easily scratched with knife. Most seashells are made of calcite. The main mineral making up limestone.

Caldera A large, circular volcanic depression the diameter of which is many times greater than that of the normal volcanic vents. Caused by explosion or collapse of a volcanic cone upon the withdrawal of magma from below.

Carbon-14 A radioactive isotope of carbon with atomic weight 14, produced by collisions between neutrons and atmospheric nitrogen. Used as a natural geologic clock to determine the age of organic material, such as wood or shells.

Cenozoic Era The geologic era of time during which mammals flourished, about 65 million years ago to the present. It consists of two periods, the Tertiary and the Quaternary.

Chert Sedimentary rock made of very fine-grained quartz. Usually made of millions of globular siliceous skeletons of tiny marine plankton called radiolarians.

Chlorite Family of green, platy, silicate minerals common in low grade metamorphic rocks

Cinder cone A volcanic cone formed by the accumulation of cinders and ash erupted from a volcanic vent.

Cirque A deep, steep-walled amphitheater caused by glacial erosion at the head of a glacier on a mountain.

Clay Particles less than 1/256 of a millimeter in diameter. Also a family of platy silicate minerals generally too small to be seen even with a microscope. A common product of rock weathering, especially of rocks containing much feldspar.

Climate The sum total of the meteorological elements that characterize the average condition of the atmosphere over a long period of time.

Coal Deposit of organic material that has been decomposed and compacted to form a black rock.

Cobble A rock fragment between the size of a baseball and a volleyball.

Columnar jointing Fracturing of lava into elongate, six-sided columns.

Composite cone A volcanic cone, usually of large dimension, built of alternating layers of lava and fragmental material.

Concretion Spherical shaped nodules formed by cementation of sand grains around local centers in sedimentary rocks.

GLOSSARY

Conglomerate Sedimentary rock made of rounded pebbles, cobbles, and boulders.

Continental crust Rocks that underlie the continents, usually granitic in composition.

Continental glacier An ice sheet covering a large part of a continent.

Contour interval The difference in elevation between two adjacent contour lines.

Contour An imaginary line on the surface of the ground, every point of which is at the same altitude.

Coral reef A reef formed by the action of reef-building coral polyps, which build internal skeletons of calcium carbonate.

Cordilleran Ice Sheet Huge ice sheet in western North America during the last Ice Age, covering much of British Columbia and the northernwestern U.S.

Crater A circular depression common at the summit of volcanoes.

Creep The slow, continuous downslope movement of rock fragments and soil.

Crevasse A deep crack in glacial ice formed by tensional stress in the ice.

Cross-bedding Inclined bedding formed by deposition of sand on the slope of a sand wave.

Cross-cutting relationships, law of Basic geologic principle that a rock body which cuts across another must be younger than the body it cuts across.

Debris flow A moving mass of water-lubricated sediment.

Deglaciation The uncovering of an area from beneath glacier ice as a result of shrinkage of a glacier.

Delta An alluvial deposit, often triangular-shaped, formed where a stream enters the ocean or a lake and drops its load of sediment.

Differential erosion Selective erosion of weaker rocks, leaving more resistant rocks standing higher.

Dike Tabular body of igneous rock formed where molten rock fills a crack in preexisting rock.

Dip The angle at which a bed or other planar feature is inclined from the horizontal.

Discharge Rate of flow exressed by volume of fluid per unit of time through a given cross-sectional area.

Divide The line of separation between drainage systems; the summit of a ridge between streams.

Drainage basin The area drained by a river system.

Drumlin A glacially smoothed, streamlined hill elongate in the direction of ice flow.

Ejecta Rock fragments and ash thrown out of a volcano.

Emergence A part of the ocean floor that has become dry land, but does not imply whether the sea receded or the land rose.

End moraine A ridge of glacial sediment deposited at the terminus of a glacier.

Eon A major subdivision of geologic time spanning billions of years.

Epidote Family of apple green to straw yellow silicate minerals common in metamorphic rocks.

Epicenter The point on the Earth's surface directly above the place of origin of an earthquake.

Epoch A subdivision of a geologic period. Example: the Pleistocene Epoch (the Ice Age)

Era A major subdivision of geologic time characterized by very different forms of life: Paleozoic Era, age of invertebrates; Mesozoic Era, age of reptiles; Cenozoic Era, age of mammals.

Erosion Wearing away of the Earth's surface by surface processes.

Erratic A boulder transported by a glacier or floating ice to a distant area.

Escarpment A cliff or steep slope.

Estuary A bay at the mouth of a river, where the tide influences the river current.

Eustatic sea level Global level of the sea.

Exposure An outcrop of rock or sediment.

Extrusive rock Rock formed by extrusion of lava on the land surface.

Facet A flat surface produced by abrasion on a rock.

Fan   A low cone-shaped accumulation of debris deposited by a stream descending from a ravine onto a plain where the material spreads out in the shape of a fan.

Fault   A fracture along which the two sides have been displaced relative to one another.

Fault block   Rocks bounded by faults.

Fault scarp   Bluff formed by movement of a fault that displaces the land surface.

Feldspar   Family of silicate minerals whose crystals are stubby prisms, generally white (plagioclase) or pink (potassium feldspar). The most common mineral of the Earth's crust.

Fiord (fjord)   A glacial trough that has been flooded by the sea.

Firn   Granular ice formed by recrystallization of snow.

Fission tracks   Microscopic tunnels in crystals and glass made by nuclear particles emitted by radioactive elements, usually uranium. The number of fission tracks in glass and zircon crystals increases with time and can be used as a dating method.

Floodplain   A strip of relatively flat land on a valley floor bordering a stream, built of sediment deposited during times of flooding.

Fluting   Smooth, deep furrows worn in the surface of rocks by glacial or stream erosion.

Fluvial   Pertaining to rivers.

Fold   Bending of rocks by tectonic forces in the Earth's crust.

Foliation   Parallel arrangement of minerals in a rock, especially platy minerals such as micas, giving the rock a look like pages in a book. Foliated rocks tend to break along the foliation and form slabs.

Garnet   Family of silicate minerals common in metamorphic rocks. Usually red but can be other colors.

Geologic column   A diagram showing the subdivisions of geologic time.

Geologic cross-section   A vertical profile of the material below the Earth's surface.

Geologic map   A map showing the distribution of rocks and sediments on the Earth's surface.

Geologic time scale A diagram showing the chronologic sequence of rocks through time

Geomorphology The study of physical and chemical processes that affect the origin and evolution of surface forms.

Glacial drift All sediment deposited directly or indirectly from a glacier or by its meltwater.

Glacial striae Scratches made by glacial erosion on smoothed surfaces of rocks.

Glacial trough U-shaped valley shaped by glacial erosion.

Glacier A body of ice, firn, and snow, originating on land and showing evidence of past or present flow.

Gneiss A light-colored, coarse grained, foliated metamorphic rock, usually granite, made by recrystallization of older rocks.

Graben A fault block down-dropped relative to the rocks on either side.

Gradient Slope expressed as the angle of inclination from the horizontal.

Granite A coarse-grained rock made of feldspar and quartz that has crystallized from molten rock at great depth below the Earth's surface or by recrystallization of older rocks.

Granitic rocks A general term for coarse grained igneous rocks composed mostly of feldspar.

Greenstone A low-grade, green metamorphic rock made by recrystallization of basalt or chemically equivalent rocks. Greenstones contain the green minerals chlorite, actinolite, and epidote, which make the rock green.

Groundwater Subsurface water.

Hanging valley. A tributary valley whose floor is noticeably higher than that of the main valley into which it opens. Generally caused by the more effective erosion of a trunk-valley glacier than its smaller tributary glacier.

Headland A projection of the land into the sea, as a peninsula or promontory.

Horn A spire-shaped peak formed by the intersecting walls of several cirques, as the Matterhorn in Switzerland.

Hornblende A specific mineral of the amphibole familiuy. Usually forms black or very dark green elongate crystals.

GLOSSARY

Hummock A small, rounded, or cone-shaped hill.

Ice sheet A large glacier of continental proportions forming a continuous cover over a land surface.

Igneous rocks Rocks formed by crystallization of molten rock.

Interglacial The time between glaciations.

Intrusion Injection of molten rock into other preexisting rocks.

Ion An atomic particle with an electrical charge.

Island arc A chain of volcanic islands, such as the Aleutian Islands.

Isostasy A state of balance among crustal rocks, analogous to floatation.

Isotope Elements with slightly different numbers of neutrons in their nucleus than is usual for a particular element and thus differ in atomic weight.

Isotope dating Measurement of the age of rocks or sediment using isotopes of radioactive minerals.

Joint A fracture in a rock with no displacement of either side other than separation.

Kame A low hill of stratified gravel and sand formed in contact with glacier ice. Commonly shows evidence of distortion as a result of collapse when supporting ice melts away.

Kame–and–kettle topography Surface formed by a kame complex interspersed with kettles.

Kame terrace A terrace of glacial sand and gravel deposited between a glacier and the valley sides.

Kettle A depression left in glacial drift by the melting of a detached mass of glacier ice that has been wholly or partly buried in the drift.

Landslide Any downhill sliding of earth.

Lateral moraine An elongate ridge of glacial sediment deposited along the sides of a glacier.

Lava Molten rock that has flowed out onto the Earth's surface.

Levee A ridge of alluvial sediment along the sides of a stream above the general level of a floodplain made by depositing of sediment overtops its channel during flooding.

GLOSSARY

Lidar Images made from airborne lasers.

Limestone A sedimentary rock composed of the mineral calcite (calcium carbonate--$CaCO_3$), commonly formed from the carbonate shells of marine creatures

Lineation Parallel arrangement of elongate minerals or groups of minerals, like pencils parallel to one another.

Lithosphere Outer shell of the Earth made of the crust and the uppermost part of the mantle.

Longshore drifting The movement of sediment parallel to the shore by longshore wave action.

Magma Molten rock formed deep in the Earth. When magma pours out on the Earth's surface it is called lava.

Magnetite Iron oxide mineral ($Fe3O4$). Usually tiny black, metallic crystals. Magnetite will attract iron.

Mantle Interior part of the Earth surrounding the core and below the crust. Made up of dense, iron-and magnesium-rich rock such as dunite and peridotite.

Marble Metamorphic rock of calcium carbonate derived from limestone by crystallization of calcite.

Marine-built terrace A bench built by deposition of sediment seaward from the shoreline.

Marine-cut terrace Flat surface made by marine erosion.

Mass wasting: The downslope movement of rock debris under the influence of gravity.

Meander A bend in the course of a stream, developed through lateral shifting of its course toward the convex side of the bend.

Melange Mixture of diverse rocks formed by multiple faulting, which brings disparate rock types together.

Mesozoic Era The age of reptiles, lasting from about 65 to 250 million years ago.

Metamorphic rocks Rocks formed by crystallization of older rocks by heat and pressure below the Earth's surface.

285

GLOSSARY

Mica Group of flat, plate-like silicate minerals, which cleave into smooth, flat flakes. **Biotite** is black. **Muscovite** is silver colored.

Mineral A naturally occurring, crystalline compound having definite chemical and physical properties

Moraine A ridge of rock debris along the margin of a glacier. An **end moraine** forms at the terminus of a glacier. A **lateral moraine** forms along the sides of a glacier.

Morainal lake Lake that forms behind a morainal dam

Mudflow A viscous flowage of mud and sediment lubricated with water.

Normal fault A steeply inclined fault along which the movement has been down the dip of the fault plane.

Oceanic rocks Rocks formed in the deep ocean.

Olivine A green, glassy mineral formed at high temperature. Common in basalt, especially oceanic basalt.

Ore deposit A mass of rock containing valuable minerals.

Outcrop An exposure of subsurface material, usually in stream banks, sea cliffs, or road cuts.

Outwash Stratified glacial deposit of sand, silt, and gravel formed downstream from a glacier by meltwater streams.

Outwash plain A topographic plain made by deposition of sand and gravel by meltwater streams from a glacier.

Paleozoic Era The age of invertebrates, spanning the time from about 600 million years ago to about 250 million years ago.

Peat An accumulation of organic material in swamps and bogs.

Phyllite A fine-grained, foliated metamorphic rock, generally derived from shale or fine-grained sandstone. Phyllites are usually black or dark gray; Differs from less crystallized slate by its sheen, which is produced by barely visible flakes of mica.

Placer Accumulation of gold or other heavy minerals in stream or beach deposits.

Plagioclase A mineral of the feldspar family commonly found in granite or lava flows, such as the Mt. Baker lavas.

Plankton Tiny, microscopic animals or plants that live floating in water.

Plate tectonics The concept of movement of crust plates that collide to form mountain ranges along continental margins.

Pleistocene The last Ice Age.

Pyrite Yellow, brassy metallic cubes of iron sulfide mineral (FeS). Known as fool's gold.

Pyroxene Family of dark green silicate minerals common in basalt and gabbro.

Quartz Glassy-looking silicon dioxide (SiO_2). One of the most common minerals in the Earth's crust, found in igneous (such as granite), metamorphic rocks (such as gneiss), sand grains in sand stone, and fracture–filled veins.

Radiocarbon Carbon-14, a radioactive isotope of carbon used to measure the age of wood and shells.

Radiocarbon dating Use of the radioactive isotope carbon-14 to determine the age of ancient wood or shells.

Radiocarbon age The age of organic material determined by the amount of the radioactive carbon isotope carbon-14. The radiocarbon age of wood or shells is slightly younger than their calendar age because of small variation in the rate of production of carbon-14.

Radiolarians Single-celled planktonic animals with skeletons of silica.

Radiolarian chert A rock made up of the siliceous shells of radiolarians.

Recessional moraine End moraine formed by a stillstand of ice during recession of a glacier.

Reef A ridge of coral and shell debris formed in warm, shallow seawater.

Relative age The age of a rock or landform compared to the age of other rocks or landforms.

Relief The difference in altitude between high and low areas.

Reverse fault A fault in which movement has occurred up the incline of a fault plane.

Ribbon chert Alternating beds of chert and thin shale resembling parallel ribbons.

Rock An aggregate of minerals.

Rock slide   A landslide in which a block of rock breaks away from the bedrock and slides along a bedding plane or fracture surface.

Sand   Grains ranging in size from 1/16th to 2 millimeters.

Sandstone   A sedimentary rock composed of sand-sized grains cemented by calcite, silica, or iron oxide.

Scarp   A cliff or steep slope. Same as escarpment.

Schist   Foliated metamorphic rock usually derived by crystallization of shale by heat and pressure.

Scoria   Volcanic rock containing many holes made by gas bubbles.

Sea cliff   A steep bluff made by wave erosion.

Sedimentary rocks   Rocks formed by deposition of grains that are later cemented together to form rock, by precipitation of chemicals in oceans or lakes, or by accumulation of shells or other organic material.

Seismic sea wave   Tsunami. Large waves generated by submarine earthquakes, slides, or volcanic eruptions.

Serpentine   Low-temperature metamorphism of minerals in ultramafic rocks to form green, greasy-looking, silicate minerals that are slippery to the touch.

Shale   Sedimentary rock form by deposition of mud on the floor of oceans or lakes or streams.

Silicate   A mineral formed from four–sided, silicon–oxygen compounds (SiO_4), the fundamental building block of silicate minerals, in which four oxygen atoms surround each silicon atom. Silicate minerals make up most of the rocks in the Earth's crust.

Silt   Grains ranging in diameter from 1/16th to 1/256th millimeter.

Siltstone   Sedimentary rock made of compacted and cemented silt grains.

Slate   A fine–grained, low grade, metamorphic rock formed by crystallization of tiny crystals of mica and chlorite from shale.

Slump   The downward slipping of a mass of rock or unconsolidated material along a concave–upward surface of failure. Common in weak, saturated, slope materials such as clay.

Soapstone   The mineral talc, a very soft, platy mineral. Can be easily carved with a knife.

Soil Decomposed rock from chemical and physical weathering and organic accumulation at the ground surface.

Spit A sandbar projecting into a body of water from the shore as a result of longshore drifting of sediment.

Stade A period during which the terminus of a glacier advanced or remained stationary.

Stagnant ice A glacier in which the ice has ceased to move.

Stamp mill A mill that crushes mining ore.

Strata Layers of sediment or sedimentary rock.

Stratification Layering of sediment or sedimentary rocks.

Striations Scratches or small grooves on rock surfaces, commonly made by glaciers dragging rock fragments over bedrock.

Strike The compass direction of a horizontal line on a bedding or fault plane.

Subduction Process of one crustal plate riding over another crustal plate as the two converge. A **subduction zone** is the area between the two plates, somewhat like a giant thrust fault.

Subglacial Beneath a glacier.

Submergence Inundation by the sea without implication as to whether the sea level rose or the land subsided.

Syncline A fold in which the beds dip inward from both sides toward the central axis.

Talc A very soft, magnesium silicate mineral, commonly called soapstone because of its softness and slippery feel.

Talus An accumulation of loose rock at the base of a cliff.

Tarn Alpine lake in the floor of a cirque caused by glacial scouring of the bedrock floor of a cirque or by damming of drainage by a moraine.

Terminal moraine A ridge of glacial deposits marking the farthest advance of a glacier.

Terrace Flat, gently inclined, or horizontal surface bordered by an escarpment. Can be either depositional or erosional. Stream terraces represent the floodplain of ancient streams when they were at a higher level. Marine terraces represent shore platforms that now stand above sea level.

Terrane A rock formation or assemblage of rock formations that share a common geologic history. A geologic terrane is distinguished from neighboring terranes by its different geologic history of formation or in its subsequent deformation and/or metamorphism.

Thrust fault A low-angle, reverse fault that pushes older rocks over younger rocks.

Till Poorly sorted, nonstratified rock debris deposited by a glacier.

Tsunami A large seismic sea wave produced by faulting of the sea floor, volcanic activity, or submarine landslides.

Tuff A fine-grained igneous rock composed of volcanic ash.

Ultramafic rock Rock very rich in pyroxene and olivine and higher in iron and magnesium than most crustal rocks. Igneous varieties are peridotite and dunite. A common metamorphic variety is serpentinite and talc.

Vein Tabular rock body made by filling of fractures with minerals precipitated from hot solutions.

Volcanic arc Arcuate chain of volcanoes formed above a subducting plate. The arc forms where a descending crustal plate becomes hot enough to cause it to melt.

Volcanic ash Volcanic rock fragments, glass, pumice, and mineral crystals

Volcanic rocks Rocks formed at the Earth's surface by the solidification of molten rock.

Water table The upper surface of the saturated zone of ground water.

Wave-built bench Gently sloping bench built by wave and current deposition of sediment.

Wave-cut bench Beveled bedrock surface produced by wave erosion.

Weathering Physical disintegration and chemical decomposition of rocks by surface processes.

REFERENCES

Armstrong, J. E., Crandell, D. R., Easterbrook, D. J., and Noble, J.A., 1965, Late Pleistocene stratigraphy and chronology in SW British Columbia and NW Washington: Geological Society of America Bulletin, vol. 76, p. 321-330.

Batchelor, C. F., 1982, Subsidence over abandoned coal mines, Bellingham, Washington: Western Washington University Master of Science thesis, 122 p.

Berger, G. W. and Easterbrook, D. J., 1993, Thermoluminescence dating tests for lacustrine, glaciomarine, and floodplain sediments from western Washington and British Columbia: Canadian Journal of Earth Sciences, vol. 30, p. 1815–1828.

Bertschi, R. G., 1992, Channel changes and flood frequency on the upper main stem of the Nooksack River, Whatcom County, Washington: Western Washington University Master of Science thesis, 111 p.

Brown, E. H., 1988, Metamorphic and structural history of the northwest Cascades, Washington and British Columbia: *in* Ernst, W. G., ed., Metamorphism and crustal evolution of the western U.S., Prentice Hall, p. 196-213.

Brown, E. H. and Gehrels, G. E., 2007, Detrital zircon constraints on terrane ages and affinities and time of orogenic events in the San Juan Island and North Cascades, Washington: Canadian Journal of Earth Sciences, vol. 44, p. 1375-1396.

Brown, E. H., Bernardi, M. L., Christenson, B. W., Cruver, J. R., Haugerud, R. A., Rady, P. M., and Sondergaard, J. N., 1981, Metamorphic facies and tectonics in part of the Cascade Range and Puget Lowland of northwestern Washington: Geological Society of America Bulletin, vol. 92, Part I, p. 170-178.

Brown, E. H. and others, 1987, Geologic map of the northwest Cascades, Washington: Geological Society of America Map and Chart Series MC-61, scale 1:100,000, 10 p.

Burke, R., 1972, Neoglaciation of Boulder Valley, Mt. Baker, Washington: Western Washington University Master of Science thesis, 47 p.

Burrows, R., 2000, Glacial chronology and paleoclimate significance of cirque moraines near Mts. Baker and Shuksan, North Cascade Range, Washington: Western Washington University Master of Science thesis.

Calkin, P., 1959, The geology of Lummi and Eliza Islands, Whatcom County, Washington: University of British Columbia Master of Science thesis, 140 p.

GLOSSARY

Cameron, V. J., 1989, The late Quaternary geomorphic history of the Sumas Valley: Simon Fraser University Master of Arts thesis, 154 p.

Carpenter, M. R., 1993, The Church Mountain sturzstrom (mega-landslides) near Glacier, Washington: Western Washington University Master of Science thesis, 71 p.

Carpenter, M. R. and Easterbrook, D. J., 1993, The Church Mountain sturzstrom (mega-landslide), Glacier, Washington: Geological Society of America Abstracts with Programs, vol. 25, no. 5, p. 18.

Carroll, P. R., 1980, Petrology and structure of the pre–Tertiary rocks of Lummi and Eliza Islands, Washington: University of Washington Master of Science thesis, 78 p.

Cary, C. M., Easterbrook, D. J., and Carpenter, M. R., 1992, Postglacial mega–landslides in the North Cascades near Mt. Baker, Washington: Geological Society of America Abstracts with Programs, vol. 24, p. 13.

Chaney, R. L., 1992, A study of gold mineralization at the Boundary Red Mountain mine, Whatcom County, Washington: Western Washington University Master of Science thesis, 108 p.

Christenson, L. G., 1986, Genesis of gold mineralization in the Lone Jack mine area, Mt. Baker mining district, Washington: Western Washington University Master of Science thesis, 87 p.

Coombs, H. A., 1939, Mount Baker, a Cascade volcano: Geological Society of America Bulletin, vol. 50, p. 1493-1509.

Crandall, R., 1983, Diatoms and magnetic anisotropy as means of distinguishing glacial till from glaciomarine drift: Western Washington University Master of Science thesis, 62 p.

Daly, R. A., 1912, North American Cordillera at the 49th Parallel: Canadian Geological Survey Memoir 38.

Danner, W. R., 1977, Paleozoic rocks of northwest Washington and adjacent parts of British Columbia: *in* Stewart, J. H., Stevens, C. H., Fritsche, A. E., eds., Paleozoic paleogeography of the western United States: Society of Economic Paleontologists and Mineralogists Pacific Section, Pacific Coast Paleogeography Symposium 1, p. 481-502.

Davidson, George, 1885, Recent volcanic activity in the United States–Eruptions of Mount Baker: Science, v. 6, p. 262.

REFERENCES

Easterbrook, D. J., 1962, Pleistocene geology of the northern part of the Puget Lowland, Washington: University of Washington Doctor of Philosophy thesis, 160 p.

Easterbrook, D. J., 1963, Late Pleistocene glacial events and relative sea-level changes in the northern Puget Lowland, Washington: Geological Society of America Bulletin, vol. 74, p. 1465-1483.

Easterbrook, D. J., 1965, Guidebook for field conference J., Pacific Northwest: VII Congress, International Association for Quaternary Research, p. 68-80.

Easterbrook, D. J., 1966, Radiocarbon chronology of late Pleistocene deposits in northwest Washington: Science, vol. 152, p. 764-767.

Easterbrook, D. J., 1966, Glaciomarine environments and the Fraser Glaciation in northwest Washington: Guidebook for First Pacific Coast Friends of the Pleistocene Field Conference, 52 p.

Easterbrook, D. J., 1969, Pleistocene chronology of the Puget Lowland and San Juan Islands, Washington: Geological Society of America Bulletin, vol. 80, p. 2273-2286.

Easterbrook, D. J., 1971, Geology and geomorphology of western Whatcom County: Western Washington University publication, 68 p.

Easterbrook, D. J., 1973, Environmental geology of western Whatcom County, Washington: Western Washington University publication, 78 p.

Easterbrook, D. J., 1974, Comparisons of late Pleistocene glacial fluctuations: *in* Quaternary glaciations in the northern hemisphere: International Geological Correlation Project 73/I/24, Report No. 1, p. 96-109.

Easterbrook, D. J., 1975, Mount Baker eruptions: Geology, vol. 3, p. 679–682.

Easterbrook, D. J., 1976, Geologic map of western Whatcom County, Washington: U.S. Geological Survey Miscellaneous Investigations Series Map I-854B.

Easterbrook, D. J., 1976, Map showing slope stability in western Whatcom County, Washington: U.S. Geological Survey Miscellaneous Investigations Series Map I–854C.

Easterbrook, D. J., 1976, Quaternary geology of the Pacific Northwest: *in* Mahany, W.C., ed., Quaternary Stratigraphy of North America, Dowden, Hutchinson, and Ross, Inc., Stroudsburg, PA., p. 441-462.

Easterbrook, D. J., 1979, The last glaciation of northwest Washington: *in* Armentrout, J. M, Cole M. R., Terbest, H., eds., Cenozoic Paleography of the Western United States,

GLOSSARY

Pacific Coast Paleogeography Symposium, Pacific Section of Society of Economic Paleontologists and Mineralogists: Los Angeles, CA, p. 177–189.

Easterbrook, D. J., 1983, Processes related to origin of debris torrents and debris chutes: Geological Society of America Abstracts with Programs, vol. 15, p. 564-565.

Easterbrook, D. J., 1986, Stratigraphy and chronology of Quaternary deposits of the Puget Lowland and Olympic Mountains of Washington and the Cascade Mountains of Washington and Oregon: *in* Sibrava, V., Bowen, D. Q., and Richmond, G. M., Quaternary Glaciations in the Northern Hemisphere, Quaternary Science Reviews, vol. 5, p. 145-169.

Easterbrook, D. J., 1988, Paleomagnetism of Quaternary sediments: *in* Easterbrook, D.J., ed., Dating Quaternary Sediments, Geological Society of America Special Paper 227, 111-122.

Easterbrook, D. J., 1992, Advance and retreat of Cordilleran ice sheets in Washington, U.S.A.: Geographie Physique et Quaternaire, vol. 46, p. 51-68.

Easterbrook, D. J., 1994, Chronology of pre-late Wisconsin Pleistocene the Puget Lowland, Washington: *in* Lasmanis, R., and Cheney, E.S., eds., Regional Geology of Washington State, Washington Division of Geology and Earth Resources, Bulletin 80, p. 191-206.

Easterbrook, D. J., 1994, Stratigraphy and chronology of early to late Pleistocene glacial and interglacial sediments in the Puget Lowland, Washington: *in* Swanson, D.A., and Haugerud, R.A., eds., Geologic Field Trips in the Pacific Northwest, Geological Society of America, p. 23-38.

Easterbrook, D. J., 1999, Surface processes and landforms: Prentice Hall, NY, 546 p.

Easterbrook, D. J., 2003, Cordilleran Ice Sheet glaciation of the Puget Lowland and Columbia Plateau and alpine glaciation of the North Cascade Range, Washington: *in* Easterbrook, D. J., ed., Quaternary Geology of the United States, INQUA 2003 Field Guide Volume, Desert Research Institute, Reno, NV, p. 265-286

Easterbrook, D. J. and Burke, R. M., 1972, Glaciation of the northern Cascades, Washington: Geological Society of America Abstracts with Programs, vol. 4, p. 152.

Easterbrook, D. J. and Kovanen, D.J., 1998, Pre-Younger Dryas resurgence of the southwestern margin of the Cordilleran Ice Sheet, British Columbia, Canada: Comments: Boreas, vol. 27, p. 225-230.

REFERENCES

Easterbrook, D. J., Kovanen, D.J., and Slaymaker, O., 2007, New developments in late Pleistocene and Holocene glaciation and volcanism in the Fraser Lowland and North Cascades, Washington: *in* Stelling, P. and Tucker, D. S. eds., Floods, Faults, and fire: Geological Field Trips in Washington State and SW British Columbia: Geological Society of America Field Guide 9, p. 31-56.

Easterbrook, D. J ., and Othberg, K. L., 1976, Paleomagnetism of Pleistocene sediments in the Puget Lowland, Washington: *in* Quaternary Glaciations of the Northern Hemisphere, International Geological Correlation Project 73/1/24, Report No. 3, p. 189-207.

Easterbrook, D. J. and Rahm, D. A., 1970, Landforms of Washington: Union Printing Co., Bellingham, WA, 156 p.

Easterbrook, D. J., Pierce, K., Gosse, J., Gillespie, A., Evenson, E., and Hamblin, K., 2003, Quaternary geology of the western United States, INQUA 2003 Field Guide Volume, Desert Research Institute, Reno, NV, p. 19-79

Easton, C. F., 1911-1931, Mt. Baker, its trails and legends: Unpublished scrapbook, Whatcom County Museum archives, Bellingham, WA.

Easton, C. F., 1920, Story of Mount Baker: Mazama, vol. 6, p. 45–53.

Franklin, R. J., 1985, Geology and mineralization of the Great Excelsior mine, Whatcom County, WA: Western Washington University Master of Science thesis, 119 p.

Frasse, F. I., 1981, Geology and structure of the western and southern margins of Twin Sisters Mountain, North Cascades, Washington: Western Washington University Master of Science thesis, 87 p.

Fuller, S. R., 1980, Neoglaciation of Avalanche Gorge and the Middle Fork Nooksack River valley, Mt. Baker, Washington: Western Washington University Master of Science thesis, 68 p.

Fuller, S. R., Easterbrook, D. J., and Burke, R. M., 1983, Holocene glacial activity in five valleys on the flanks of Mt. Baker, Washington: Geological Society of America Abstracts with Programs, vol. 15, p. 430–431.

Gannaway W. and Holsather, K., 2004, Whatcom then and now: LoneJack Mountain Press, Bellingham, WA, 137 p.

Glover, S. L., 1935, Oil and gas possibilities of western Whatcom County: Washington Division of Geology Report of Investigations, vol. 2, 69 p.

GLOSSARY

Gowan, M. E., 1989, The mechanisms of landslide initiation and flood generation in the Boulder Creek basin, Whatcom County, Washington: Western Washington University Master of Science thesis, 189 p.

Gower, H. D., 1978, Tectonic map of the Puget Sound region, Washington, showing locations of faults, principal folds, and large-scale Quaternary deformation: U.S. Geological Survey Open–File Report 78–426, 22 p.

Griggs, P. H., 1970, Palynological interpretation of the type section, Chuckanut Formation, northwestern Washington: in Kosanke, R. M. and Cross, A. T., eds., Symposium on palynology of the Late Cretaceous and early Tertiary: Geological Society of America Special Paper 127, p. 169-212.

Hansen, B. S. and Easterbrook, D. J., 1974, Stratigraphy and palynology of late Quaternary sediments in the Puget Lowland, Washington: Geological Society of America Bulletin, vol. 85, p. 587–602.

Harper, J. T., 1992, The dynamic response of glacier termini to climatic variation during the period 1940-1990 on Mt. Baker, Washington, U.S.A.: Western Washington University Master of Science thesis, 132 p.

Harrison, A. E., 1961, Fluctuations of the Coleman Glacier, Mt. Baker, Washington: Journal of Geophysical Research, vol. 66, p. 649-650.

Harrison, A. E., 1961, Ice thickness variations at an advancing front, Coleman Glacier, Mt. Baker, Washington: Journal of Glaciology, vol. 3, p. 1168-1170.

Harrison, A. E., 1970, Fluctuations of Coleman Glacier, Mt. Baker, Washington, U.S.A.: Journal of Glaciology, vol. 9, p. 393–396.

Hagerud, R. A., Brown, E. H., Tabor, R. W., Kriens, B. J., and Mcgroder, M. F., 1994, Late Cretaceous and early Tertiary orogeny in the North Cascades: in Swanson, D. A. and Haugerud, R. A., eds., Geologic Field Trips in the Pacific NW, Geological Society of America Guidebook, p. 2e1-2e51.

Heikkinen, O. O., 1983, Climatic changes during recent centuries as indicated by dendrochronological studies, Mt. Baker, Washington, U.S.A: in Morner, N. A. and Karlen, W., eds., Climatic changes on a yearly to millennial basis–Geological, historical, and instrumental records: Reidel Publishing Company, p. 353-361.

Heikkinen, O. O., 1984, Dendrochronological evidence of variations of Coleman Glacier, Mt. Baker, Washington, U.S.A.: Arctic and Alpine Research, vol. 16, p. 53-64.

REFERENCES

Hildreth, W., Fierstein, J., and Lanphere, M., 2003, Eruptive history and geochronology of the Mt. Baker volcanic field, Washington: Geological Society of America Bulletin, vol. 115, p. 729-764.

Hillhouse, D. N., 1956, Geology of the Vedder Mountain–Silver Lake area: University of British Columbia Master of Science thesis, 52 p.

Holsather, K. and Gannaway, W., 2008, Bellingham, then and now: LoneJack Mountain Press, Bellingham, WA, 221 p.

Hyde, J. H. and Crandell, D. R., 1978, Postglacial volcanic deposits at Mt. Baker, Washington, and potential hazards from future eruptions: U.S. Geological Survey Professional Paper 1022–C, 17 p.

Impero, M. G., 2007, The Lone Jack, king of the Mt. Baker Mining district: 180 p.

James, E. W., 1980, Geology and petrology of the Lake Ann stock and associated rocks: Western Washington University Master of Science thesis, 57 p.

Jeffcott, P. R., 1963, Chechaco and sourdough--Being an account of the hectic pursuit of gold in the Mt. Baker mining district of Whatcom County, Washington, 1858-1960: Pioneer Printing Company, Bellingham, WA, 181 p.

Jenkins, O. P., 1923, Geological investigation of the coal fields of western Whatcom County, Washington: Washington Division of Geology Bulletin 28, 135 p..

Johnson, S. Y., 1984, Cyclic fluvial sedimentation in a rapidly subsiding basin, northwest Washington: Sedimentary Geology, vol. 38, p. 361-391.

Johnson, S. Y., 1984, Stratigraphy, age, and paleogeography of the Eocene Chuckanut Formation, northwest Washington: Canadian Journal of Earth Sciences, vol. 21, p. 92–106.

Jones, J. T., 1984, The geology and structure of the Canyon Creek–Church Mountain area, North Cascades, Washington: Western Washington University Master of Science thesis, 125 p.

Jones, R. F., 1958, Boundary town: Fleet Printing Co., Vancouver, WA, 306 p.

Kelly, J. M., 1970, Mineralogy and petrology of the basal Chuckanut Formation in the vicinity of Lake Samish, Washington: Western Washington State College Master of Science thesis, 81 p.

Kiver, E. P., 1974, The summit firn caves of Mt. Baker: International Glaciospeleological Survey Bulletin 3, p. 5–85.

GLOSSARY

Kiver, E. P., 1975, The first exploration of Mount Baker ice caves: Explorers Journal, vol. 53, p. 84–87.

Kiver, E. P., 1978, Mount Baker's changing fumaroles: Ore Bin, vol. 40, p. 133-145.

Koert, D. and Biery, G., 1980, Looking back: Lynden Tribune, 248 p.

Kovanen, D. J. and Easterbrook, D. J., 1996, Extensive readvance of late Pleistocene (Y.D.?) alpine glaciers in the Nooksack River Valley, 10,000 to 12,000 years ago, following retreat of the Cordilleran Ice Sheet, North Cascades, Washington: Friends of the Pleistocene, Pacific Coast Cell Field Trip Guidebook, 74 p.

Kovanen, D. J. and Easterbrook, D. J. 2001, Late Pleistocene, post-Vashon, alpine glaciation of the Nooksack drainage, North Cascades, Washington: Geological Society of America Bulletin, vol. 113, p. 274–288.

Kovanen, D.J. and Easterbrook, D.J., 2002, Paleodeviations of radiocarbon marine reservoir values for the NE Pacific: Geology, vol. 30, p. 243-246.

Kovanen, D.J. and Easterbrook, D.J., 2002, Timing and extent of Allerod and Younger Dryas age (ca. 12,500–10,000 ^{14}C yr B.P.) oscillation of the Cordilleran Ice Sheet in the Fraser Lowland, western North America: Quaternary Research, vol. 57, p. 208-224.

Kovanen, D.J., Haugerud, R.A., and Easterbrook, D.J., in press, Geomorphic map of western Whatcom County, Northwest Washington: U.S. Geological Survey.

Krom, M. M., 1937, The Boundary Red Mountain mine, Whatcom County, Washington: University of Washington Bachelor of Science thesis, 135 p.

Leiggi, P. A., 1986, Structure and petrology along a segment of the Shuksan thrust fault, Mt. Shuksan area, Washington: Western Washington University Master of Science thesis, 207 p.

Lindstrom, P. M., 1941, The Lone Jack gold mine, Whatcom County: University of Washington Bachelor of Science thesis, 71 p.

Liszak, J. L., 1982, The Chilliwack Group on Black Mountain: Western Washington University Master of Science thesis, 104 p.

Majors, H. M., 1978, Mt. Baker: A chronicle of its historic eruptions and first ascent: Northwest Press, Seattle, 226 p.

REFERENCES

Malone, S. D. and Frank, D. G. 1975, Increased heat emission from Mount Baker, Washington: Eos (American Geophysical Union Transactions), vol. 56, no. 10, p. 679-685.

Mattinson, J. M., 1972, Ages of zircons from the northern Cascade Mountains, Washington: Geological Society of America Bulletin, vol. 83, p. 3769–3783.

McKeever, D., 1977, Volcanology and geochemistry of the south flank of Mount Baker, Cascade Range, Washington: Western Washington University Master of Science thesis, 126 p.

McLellan, R. D., 1927, The geology of the San Juan Islands: University of Washington Publications in Geology, vol. 2, 185 p.

Miller, G. M., 1979, Western extent of the Shuksan and Church Mountain thrust plates in Whatcom, Skagit, and Snohomish Counties, Washington: Northwest Science, vol. 53, p. 229-241.

Misch, P., 1952, Geology of the North Cascades of Washington: The Mountaineer, vol. 45, p. 4-22.

Misch, P., 1966, Tectonic evolution of the northern Cascades of Washington–A west-cordilleran case history: in Canadian Institute of Mining and Metallurgy, Symposium on the tectonic history and mineral deposits of the western Cordillera: Canadian Institute of Mining and Metallurgy Special Volume 8, p. 101-148.

Misch, P., 1977, Bedrock geology of the north Cascades: in Brown, E. H. and Ellis, R. C., eds., Geological excursions in the Pacific Northwest, Geological Society of America, p. 1-62.

Misch, P., 1988, Tectonic and metamorphic evolution of the North Cascades: An overview: in Ernst, W. G., ed., Metamorphism and crustal evolution of the Western U.S., Prentice Hall, p. 179-195.

Misch, P., 2008, Geology of the Northern Cascades: Cascade metamorphic suite: Unpublished manuscript available on Western Washington University Dept. of Geology website, 91 p.

Moen, W. S., 1962, Geology and mineral deposits of the north half of the Van Zandt quadrangle, Whatcom County, Washington: Washington Division of Mines and Geology Bulletin 50, 129 p.

Moen, W. S., 1969, Mines and mineral deposits of Whatcom County, Washington: Washington Division of Mines and Geology Bulletin 57, 134 p.

GLOSSARY

Mustoe, G. E., 1982, The origin of honeycomb weathering: Geological Society of America Bulletin, vol. 93, p. 108–115.

Mustoe, G. E. and Pevear, D. R., 1983, Vertebrate fossils from the Chuckanut Formation of northwest Washington: Northwest Science, vol. 57, p. 119-124.

Newcomb, R. C., Sceva, J. E., and Stromme, O., 1949, Ground–water resources of western Whatcom County, WA: U.S. Geological Survey Open-File Report 50-7, 134 p.

Onyeagocha, A. C., 1978, Twin Sisters dunite--Petrology and mineral chemistry: Geological Society of America Bulletin, vol. 89, p. 1459-1474.

Pabst, M. B., 1968, The flora of the Chuckanut Formation of northwestern Washington– The Equisetales, Filicales, Coniferales: University of California Publications in Geological Sciences, vol. 76, 85 p.

Ragan, D. M., 1961, The geology of the Twin Sisters dunite in the northern Cascades, Washington: University of Washington PhD thesis, 88 p.

Ragan, D. M., 1963, Emplacement of the Twin Sisters Dunite, Washington: American Journal of Science,

Ragan, D. M., 1967, The Twin Sisters dunite, Washington: *in* Wyllie, P. J., ed., Ultramafic and related rocks: John Wiley and Sons, p. 160-167.

Rigg, G. B., 1958, Peat resources of Washington: Washington Division of Mines and Geology, Bulletin 44, 272 p.

Russell, I. C., 1893, Geology of the Cascade mountains in northern Washington: U.S. Geological Survey 20th Annual Report, Part 2, p. 83-310.

Schasse, H. W., 1988, Coal exploration in Whatcom County: Washington Geologic Newsletter, vol. 16, p. 11.

Schmierer, A. C., 1983, Northing up the Nooksack: Pacific NW National Parks and Forests Association, 79 p.

Sevigny, James H., 1983, Structure and petrology of the Tomyhoi Peak area, north Cascade Range, Washington: Western Washington University Master of Science thesis, 203 p.

Smith, G. O. and Calkins, F. C., 1904, A geological reconnaissance across the Cascade Range near the 49th Parallel: U. S. Geological Survey Bulletin, vol. 325, 103 p.

REFERENCES

Sondergaard, J. N., 1979, Stratigraphy and petrology of the Nooksack Group in the Glacier Creek-Skyline Divide area, north Cascades, Washington: Western Washington University Master of Science thesis, 103 p.

Stavert, L. W., 1971, A geochemical reconnaissance investigation of Mount Baker andesite: Western Washington University Master of Science thesis, 60 p.

Strangberg, H. V., 1937, Glacier recession on Mount Baker–1937: Mountaineer, vol. 30, p. 33.

Stuiver, M. and Grootes, P.M.,.2000, GISP2 Oxygen Isotope Ratios: Quaternary Research, 54/3.

Swan, V. L., 1980, The petrogenesis of the Mount Baker volcanics, Washington: Washington State University PhD thesis, 652 p.

Syverson, T. L., 1984, History and origin of debris torrents in the Smith Creek drainage, Whatcom County, Washington: Western Washington University Master of Science thesis, 84 p.

Tabor, R. W. and Haugerud, R., 1999, Geology of the North Cascades: The Mountaineers, 143 p.

Tabor, R. W., Haugerud, R. A., and Miller, R. B., 1989, Overview of the geology of the North Cascades: in American Geophysical Union, International Geological Congress Field Trip, 62 p.

Tabor, R. W., Haugerud, R. A., Hildreth, W., and Brown, E. H., 2004, Geologic map of the Mt. Baker 30 x 60-minute quadrangle, Washington: U.S. Geological Survey Map I-2660.

Tepper, J. H., 1985, Petrology of the Chilliwack composite batholith, Mt. Sefrit area, North Cascades, Washington: Univ. of Washington Master of Science thesis, 102 p.

Van Siclen, Carla Cary, 1994, Geologic, hydrologic, and climatic factors influencing Glacier Creek basin, Whatcom County, Washington: Western Washington University Master of Science thesis, 271 p.

Webber, S. J., 2001, Late Pleistocene littoral deposits in the Deming sand at Bellingham Bay, and their implications for relative sea level changes: Washington: Western Washington University Master of Science thesis, 158 p.

Westgate, J. A., Easterbrook, D. J., Naeser, N. D., and Carson, R. J., 1987, Lake Tapps tephra: An early Pleistocene stratigraphic marker in the Puget Lowland, Washington: Quaternary Research, vol. 28, p. 340-355.

GLOSSARY

Whitney, J. D., 1889, *in* Mt. Baker, its trails and legends: Unpublished manuscript by the Mt. Baker Club in Whatcom County Museum archives, Bellingham, Washington.

Wolff, F. E., Brookshier, M. I., and Norman, D. K., 2008, Inactive and abandoned mine lands–Boundary Red Mountain Mine, Mt. Baker Mining District, Whatcom County, Washington: Washington Division of Geology and Earth Resources, Information Circular 99, 7 p.

Wolff, F. E., McKay, D. T., Brookshier, M. I., and Norman, D. K., 2005, Inactive and abandoned mine lands–Lone Jack Mine, Mt. Baker Mining District, Whatcom County, Washington: Washington Division of Geology and Earth Resources, Information Circular 98, 9 p.

Wolff, F. E., McKay, D. T., Norman, D. K., and Brookshier, M. I., 2004, Inactive and abandoned mine lands–Great Excelsior Mine, Mt. Baker Mining District, Whatcom County, Washington: Washington Division of Geology and Earth Resources, Information Circular 98, 10 p.

Woodruff, Elmer Grant, 1914, The Glacier coal field, Whatcom County, Washington: U.S. Geological Survey Bulletin 541, p. 389-398.

Ziegler, Charles B., 1985, The structure and petrology of the Swift Creek area, western North Cascades, Washington: Western Washington University Master of Science thesis, 191 p.

Zobrist, E., 1979, Ghost towns of Lake Whatcom: 95 p.

GUIDE TO THE GEOLOGY
ALONG CHUCKANUT DRIVE

Cum miles	Miles since last point	Descripion
0	0	Fairhaven Park, south Bellingham. Outcrop of Chuckanut sandstone on right.
0.6	0.6	Roadcut on right Everson glaciomarine pebbly clay deposited from floating ice in marine water 11,700 yrs ago.
2.1	1.5.	Chuckanut sandstone dipping 65°S, strike N65°W, on the SW limb of the Chuckanut anticline, which plunges northwesterly..
4.7	2.6	Chuckanut sandstone and shale near the axis of the Chuckanut syncline, strike N54°W, dip 60°N.
5.0	0.3	Chuckanut sandstone and shale, strike N64°W, dip 47°N Cross-bedding is well developed in sandstone on the SW limb of the Chuckanut syncline. Outcrops from here to location 14 are on the same syncline.
5.3	0.3	Chuckanut sandstone, strike N75°W, dip 30°N
5.4	0.1	Larrabee State Park boundary.
5.6	0.2	North entrance to Larrabee State Park
5.7	0.1	South Entrance to Larrabee State Park. Parking area on right. Chuckanut sandstone and interbeds of shale here dip 85°S to nearly vertical, strike N70°W
5.9	0.2	Whatcom-Skagit county boundary
6.3	0.4	Roadside park. Leaf fossils occur in abundance in Chuckanut shale. Strike N70°W, dip 65°N.
7.2	0.9	Parking area.
7.4	0.2	Cliffs of Chuckanut sandstone, shale, and coal dipping to the north.
7.7	0.3	Roadside park. Shale, interbedded with sandstone, contains plant fossils. Strike N75°W, dip 56°N.
7.9	0.2	Chuckanut sandstone and shale; strike N74°W, dip 56°N. Fossil palm fronds 2-3 feet in diameter are exposed on a bedding plane and are visible from the road. Other fossil leaves are abundant in carbonaceous shale beneath the fossil palms. DO NOT TRY TO CHIP OUT THE PALM FRONDS, they are easily destroyed.

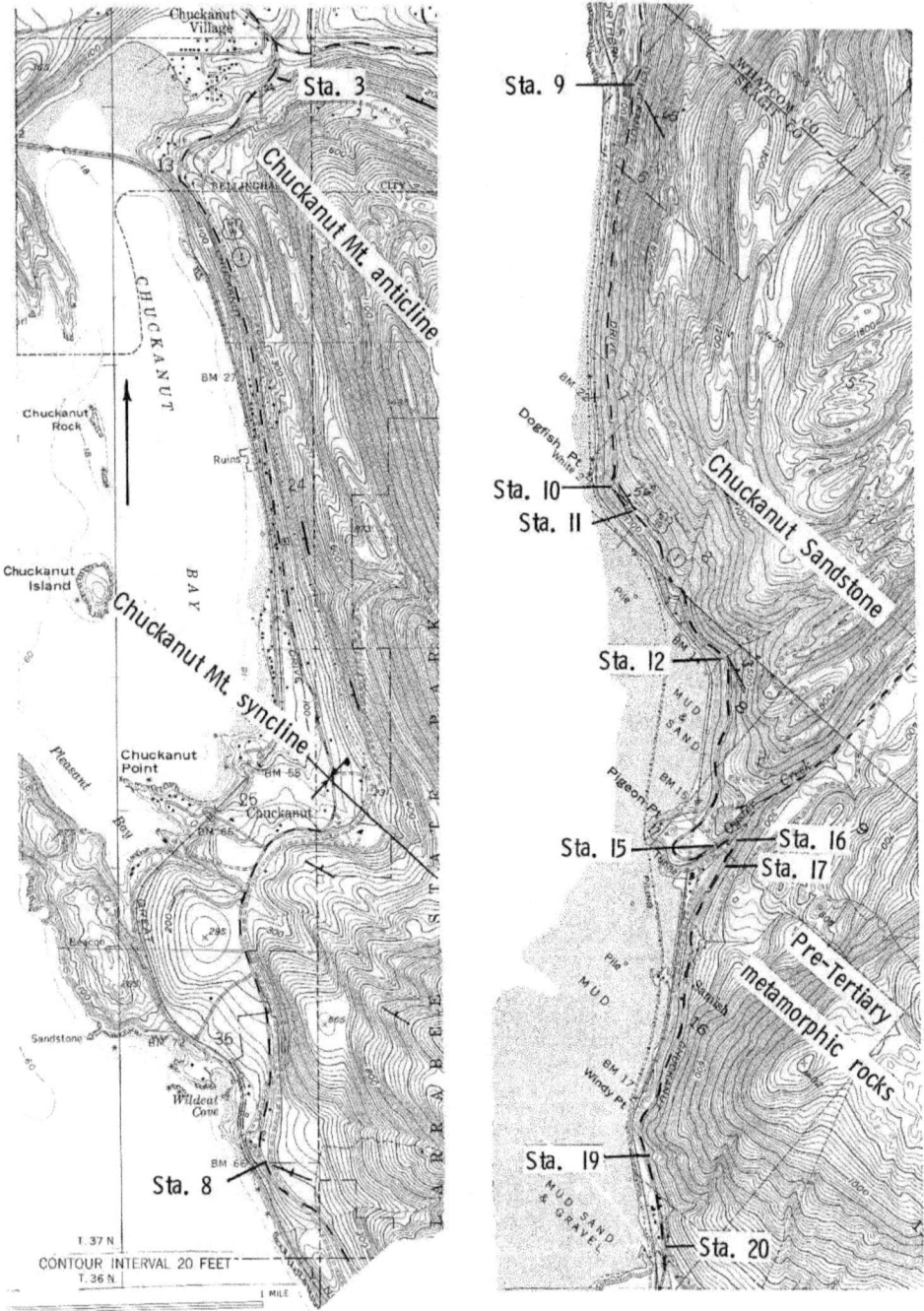

MAP OF CHUCKANUT DRIVE.

Figure 299. Map of Chuckanut Drive showing points of geologic interest.

304

8.2	0.3	Parking area. Chuckanut sandstone, strike N68°W, dip 61°N.
8.5	0.3	Chuckanut sandstone, shale, and coal in cliffs to the left. A local synclinal flexure occurs near the bend in the road.
8.8	0.3	Roadside park. Outcrops of Chuckanut sandstone; strike N69°W, dip 42°N.
9.1	0.3	Cross approximate location of the contact between the Chuckanut Formation and underlying Darrington phyllite and greenstones. The contact is not exposed here.
9.3	0.2	From Oyster Creek to the parking area are outcrops of altered greenstones. Sheared rock in fault zones up to 6" wide occur in places.
9.6	0.3	Oyster Creek. Blue–green, schistose serpentine at the south end of bridge.
9.9	0.5	Outcrops of Darrington phyllite. Foliation trends N38°W, dips 66°N. The phyllite occurs at numerous places on the left.
10.4	0.5	Park on right. Silvery-gray, well foliated, Darrington phyllite. Microscopic lenses of granular quartz interfingering with elongate stringers of fine-grained mica give the rock a well-defined foliation, which strikes N40°W, dips 30°N. Accessory minerals include epidote and graphite.
10.8	0.4	Quarry on left. Serpentinized greenstone. The serpentine is in places massive, elsewhere occurring on slickensided surfaces on incompletely serpentinized parent rock.
11.1	0.3	Outcrop on left side of road. Silvery-gray to black, fine-grained, foliated Darrington phyllite, in places containing graphite and cut by lenses of quartz.

GUIDE TO THE GEOLOGY ALONG INTERSTATE 5 FROM BURLINGTON TO BELLINGHAM

Cum miles	Miles since last point	Description
0	0	Samish River, leave floodplain of Skagit River and ascend a late Pleistocene outwash terrace and delta about 11,500 years old.
0.4	0.4	Outwash channel and terraces. The Samish River flows on a broad valley floor about 1/2 mile wide with meander bends greater than about one mile. The present meanders of the Samish River are only about 1/10th of a mile, indicating that the Samish River is flowing in a valley previously made by a much larger stream.
		A prominent terrace occurs at an elevation of about 100 feet at the junction of the valley with the present Skagit delta. Pebble to cobble gravel which make up the terrace near its southern terminus is exposed in gravel pits just west of Interstate 5 and east of Highway 99. The present Samish River drainage is almost entirely in phyllite, so the many granitic pebbles and cobbles suggest that the river which deposited the gravel was a meltwater stream carrying material from British Columbia. Conspicuous pebbles and cobbles of Mt. Baker andesite in the gravel suggest contributions of sediment by the Nooksack River drainage.
		About a mile to the north, the continuation of the terrace appears to be a cut terrace, whereas the surface here appears to be essentially a fill terrace banked against older glacial deposits. Two miles to the east, an abandoned channel and terrace correlate with this terrace. Just to the north, the land surface rises rather sharply where the terrace is banked against Everson glaciomarine stony clay. Thus, the terrace is younger than the 11,700-year-old Everson glaciomarine sediments.
		When the Cordilleran Ice Sheet of the Sumas Stade blocked the Nooksack River valley at Deming about 11,400 radiocarbon years ago, glacial meltwater flowed southward in the Nooksack South Fork valley into the upper Samish River valley and built a delta into the Skagit Valley about 100 feet above present sea level. The meltwater channel is cut into Everson glaciomarine stony clay and the channel truncates multiple shorelines between 100 and 250 feet above present sea level, showing that the channel is younger than these features.

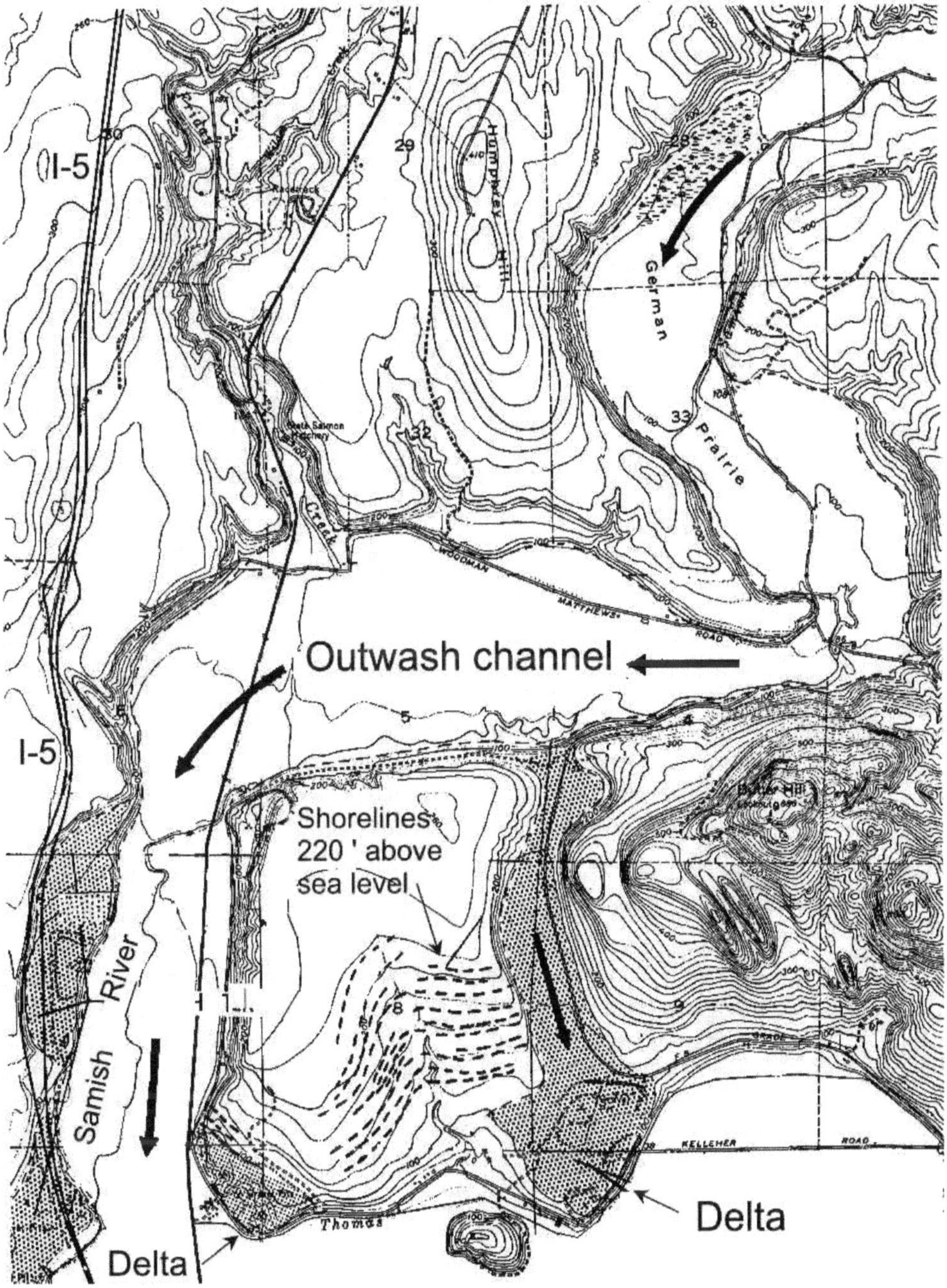

Figure 300. The Samish River outwash channel cutting ancient shorelines and into the Bellingham stony clay (11,700 yrs old). Outwash terraces along the valley sides terminate at deltas just north of the Skagit River floodplain. (Easterbrook, 1994)

1.1	0.7	Leave the outwash terrace and ascend onto Everson glaciomarine stony clay. Roadcuts are in glaciomarine sediments against which the outwash terrace has been built.
2.3	1.2	Bow Hill Road overpass. Everson glaciomarine stony clay exposed in road cuts since the last point. Take the freeway exit on the right. Exposed in the road cut at the top of the hill is about 10 feet of Everson glaciomarine pebbly silt overlying Vashon till. The glaciomarine sediments consists mostly of silt and fine material with scattered pebbles. The meltwater channel of the Samish River valley, which is cut into the glaciomarine stony clay, may be seen to the southeast. More roadcuts in glaciomarine stony clay occur on the west side of the freeway. Return to the freeway via the onramp.
6.3	4.0	Outcrop of Darrington phyllite on the right. This low–grade metamorphic rock contains many lenses of quartz, which contrast with the dark color of the phyllite.
6.9	0.6	Alger exit. A drumlin on the right has been cut by the road to Alger. To look at a cross section of the sediments in the drumlin, take the exit and stop at the first road cuts.
7.9	1.0	North end of a drumlin, cut lengthwise by the freeway. Streamlining of the hill was made by the Cordilleran Ice Sheet.
8.6	0.7	Entering Whatcom County. The foothills to the east are made of Chuckanut sandstone lying on Darrington phyllite.
9.6	1.0	Darrington phyllite exposed in road cuts.
10.6	1.0	Contact between Chuckanut sandstone and Darrington phyllite and greenstones. Beneath the Chuckanut Formation, well-foliated greenschist contains many elongate quartz pods and lenses which are oriented parallel to the foliation. Outcrops of greenschist occur for about 120 feet south of the contact with Chuckanut sandstone, but the next 150 feet of outcrop is graphitic phyllite, then greenschist continues again to the south. The contact between greenschist and phyllite is sharply defined and foliation is not continuous across the contact. Ground-up rock, suggestive of faulting, is present at the contact between the phyllite and greenschist. The Darrington phyllite and greenschist are overlain by Chuckanut sandstone. A basal conglomerate, consisting of quartz pebbles weathered and eroded from quartz pods in the underlying Darrington phyllite, occurs near the contact and dips 60° to the north. ,The Chuckanut sandstone above the conglomerate is interbedded with organic shale (dip 73°N, strike N63°E). Outcrops of north-dipping Chuckanut sandstone extend along the road for 0.3 miles.

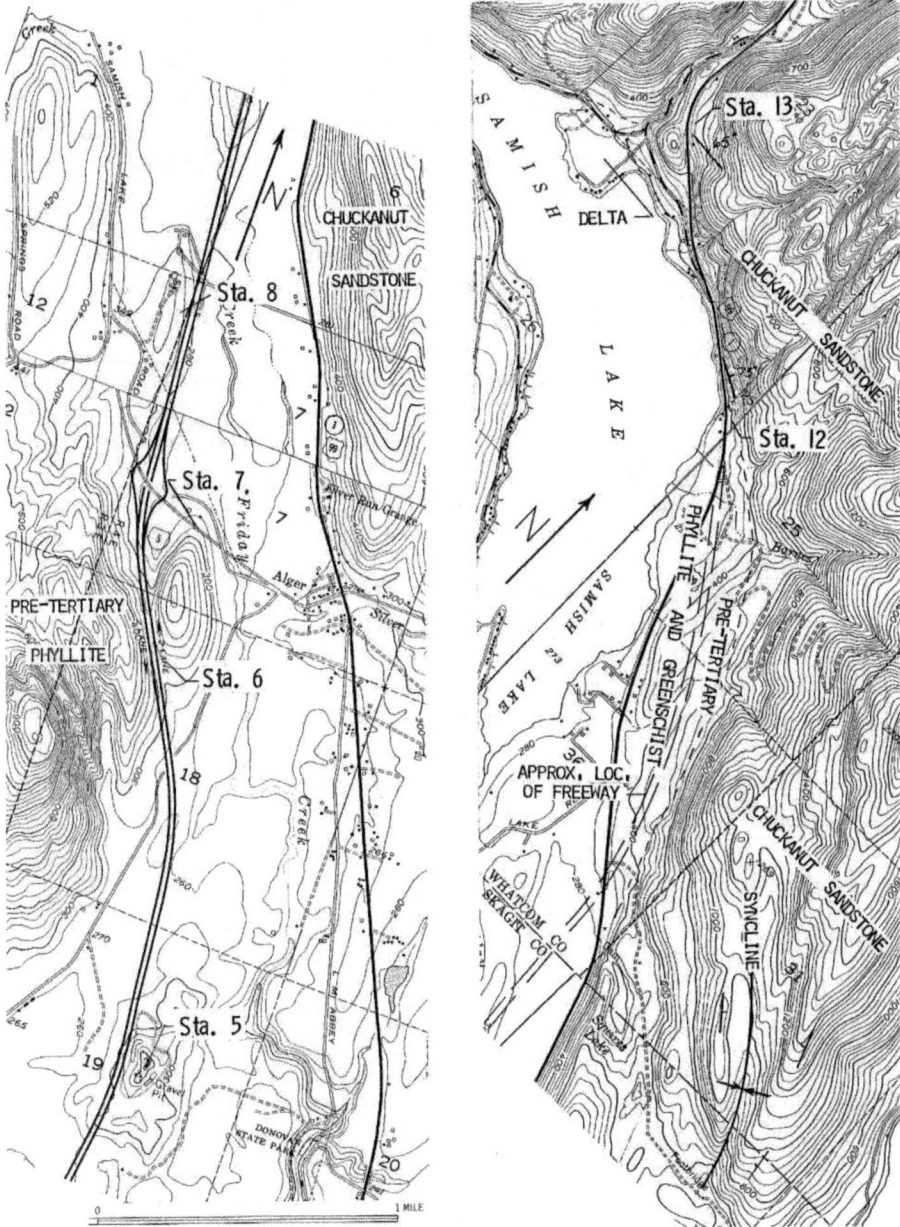

TOPOGRAPHIC MAP OF THE AREA ADJACENT TO THE SAMISH FREEWAY.

43

Figure 301. Map of points of geologic interest along Interstate 5.

11.4	0.5	Chuckanut sandstone dipping 65 °N, strike N85°E.
12.3	0.9	Samish Lake overpass.
12.8	0.5	Glacial erratics on right side of road were carried south from British Columbia by ice and deposited during the last Ice Age. The composition of the boulder is different than the bedrock upon which it rests, indicating a source in Canada.
13.5	0.7	Chuckanut sandstone and shale, dip 53°N, strike N72°W. The beds dip away from the freeway but joints (fractures) dip toward the freeway. From this point westward for 3 miles a bad slide area has developed from large blocks of rock sliding down joint planes that have been undercut by road excavations.
14.0	0.5	Chuckanut sandstone, dipping 74°N, strike N70°W on the northeast flank of the Chuckanut anticline.
16.4	2.4	Chuckanut sandstone and shale for last 2.4 mi. Wood–bearing conglomerate here dips 81°N, strike N60W.
18.2	1.8.	Take Fielding Street exit. Turn left at the stop sign and continue to the second stop light. Turn right and drive to the stop sign. Turn right on Donovan Ave and drive to the end of the street.
		Donovan erratic. At the end of the last Ice Age, a remnant of the Cordilleran Ice Sheet strewed boulders of the Jackass Mt. conglomerate from the Fraser Valley all over Whatcom County. The huge erratic boulder at the end of Donovan Ave. was one of the largest and a must–see geology field trip stop for many years. When the freeway was built in the early 1960s, highway engineers, despite numerous pleas to leave as a landmark, blew the huge boulder apart and pushed the remaining fragments into the pile that now remains.

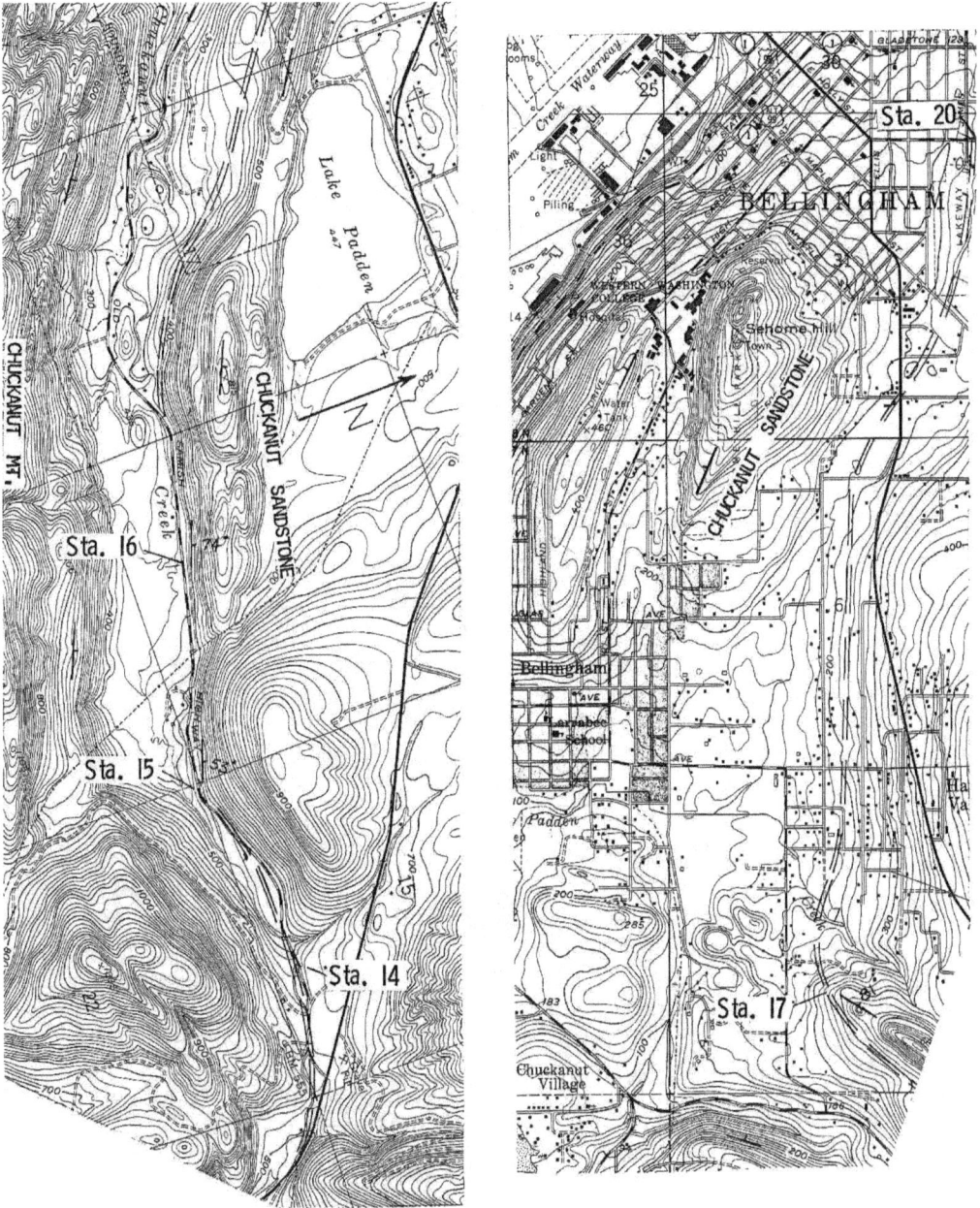

TOPOGRAPHIC MAP OF THE AREA ADJACENT TO THE SAMISH FREEWAY.

Figure 302. Map of points of geologic interest along Interstate 5.

311

GUIDE TO THE GEOLOGY ALONG THE MT. BAKER HIGHWAY FROM BELLINGHAM TO ARTISTS POINT

Cum miles	Miles since last point	Description	
0	0	Stop light on east side Sunset Drive/Interstate 5 overpass	
0.8	0.8	Intersection with Hannegan road. Turn left.	
1.1	0.3	The road drops down into the Squalicum outwash channel eroded into Bellingham glaciomarine stony clay by meltwater from a remnant of the Cordilleran Ice Sheet whose margin was a few miles to the north. Because the channel cuts into Bellingham stony clay, it must be younger than 11,700 years.	
1.7	0.6	Intersection with Bakerview road	
2.7	1.0	Intersection with Van Wyck road	
2.9	0.2	Ascend the Squalicum end moraine. The moraine is older than a peat bog in Tenmile channel a few miles to the north, dated at 11,100 years and rests on 11,700 year-old Bellingham stony clay. Thus, the age of the moraine is between 11,700 and 11,100 years.	
3.8	0.9	Turn around	
4.8	1.0	Intersection with the Van Wyck road. Turn left.	
5.8	1.0	Intersection with the Dewey road. Descend into the upper part of the Squalicum outwash channel. Remnants of the moraine rest on a terrace within the channel on the north (left) side of the road. This means that most of the channel must be older than the moraine in order for it to drape down into it. Thus, the original channel was cut by meltwater from ice to the north, then the glacier advanced, covering this part of the channel and depositing the moraine across the channel. The glacier then receded to the north and meltwater cut the channel slightly deeper, leaving the moraine remnant on a terrace that was the valley floor when the ice was there.	
6.9	1.1	Intersection with Mt. Baker Highway. Turn left.	7.8
7.8	0.9	Pass Everson Goshen road.	
8.8	1.0	Intersection with Mission road. Roadcut approaching the Y-Road exposes silty, sandy, pebbly glacial till making up the Squalicum moraine.	

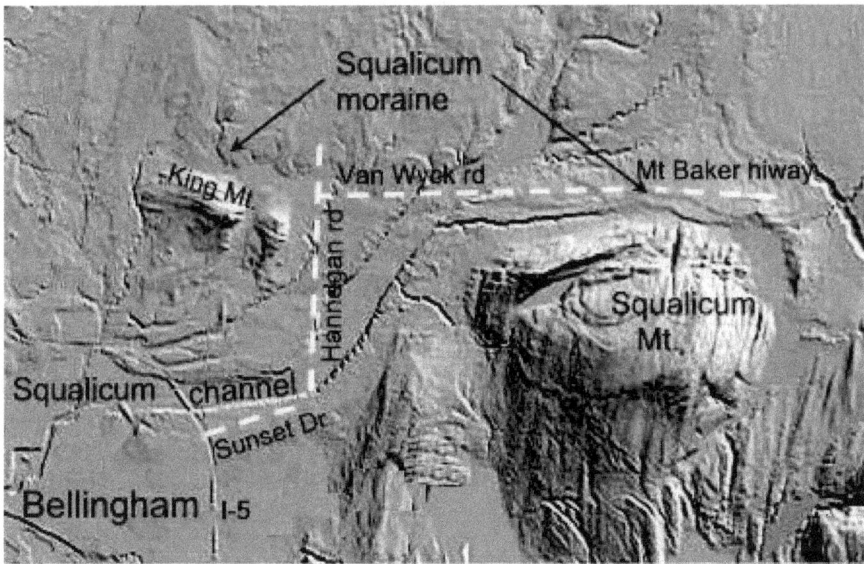

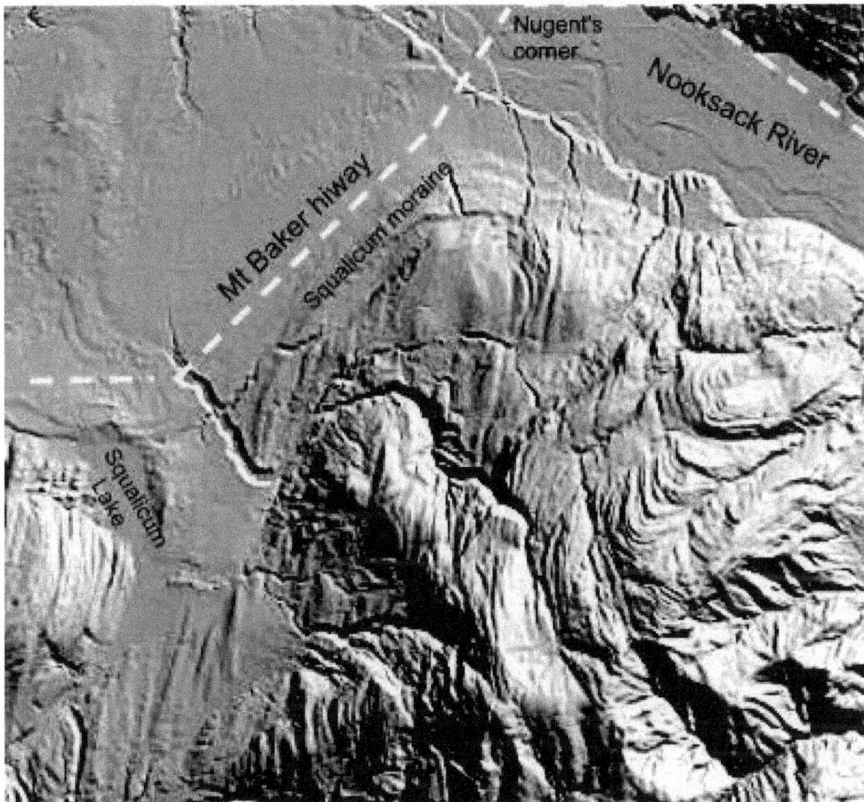

Figure 303. Lidar images of field trip route. White dashed lines mark the route. (Modified from USGS lidar image provided by Whatcom County)

9.0	0.2	Turn into parking area just before intersection with Y-Rd
		Walk south on the trail to Squalicum Lake, which occupies a depression left by melting out of a block of glacial ice. Prior to deposition of the moraine along the Mt. Baker Highway, the glacier extended about a mile south of Squalicum Lake where it deposited a moraine across the valley. When the ice melted back to the position of the highway moraine, it left a deep depression in which Squalicum Lake formed. The bottom of a peat bog that surrounds the lake gave a radiocarbon date of 10,300 years. but the depression is thought to be at least 11,000 years old because of radiocarbon dates from bogs farther north.
9.5	0.5	Cross Anderson Creek.
10.4	0.9	Intersection with the Kelly Road. Turn right and drive to the top of the hill where a road cut exposes sediments in one of the multiple moraines here. Return to Mt. Baker Highway, turn right.
11.8	1.4	Intersection with the Smith Road. Turn right and drive to the end of the road.
		Type locality of the Everson glaciomarine stony clay in bluffs along the Nooksack River. This is private property so be sure to ask permission to cross the pasture to the bluffs. The bluffs here are on the outside of a bend of the Nooksack River and have been undercut for many years, causing them to recede at a rapid rate. The glaciomarine stony clay here becomes saturated during the rainy season, making the clay highly unstable, so do not attempt to go onto the bluffs in the winter wet season or after a heavy rain. For a description of the sediments exposed here see p.175-179.
		Return to Mt. Baker Highway. Turn right.
12.2	0.4	Roadcut descending to Cedarville. Bellingham stony clay overlain by thin, fluvial sand of a stream terrace.
12.6	0.4	Cross Nooksack River.
12.9	0.3	Nugents corner
17.0	6.1	Deming
17.1	0.1	Outcrops of 50 million–year–old Chuckanut sandstone. Abandoned quarry on the left side of road in sandstone, black, carbon-rich shale, and conglomerate. At the far end of the quarry, an upright fossil tree about 10 feet high in massive sandstone with roots in underlying shale could formerly be seen in the quarry wall, along with . a smaller tree trunk. Fossil leaves were abundant in the black shale. Unfortunately, quarrying has now destroyed the tree and it can no longer be seen.

17.3	0.2	Intersection with Highway 9. The roadcut on the left (north) side of the highway is now totally overgrown with vegetation, but in 1960, it was nicely exposed. Most of the roadcut was composed of Bellingham glaciomarine stony clay containing marine fossils, mostly *Nuculana*, near the top of the exposure. The glaciomarine stony clay is underlain by bedded Deming sand in the lower part of the cut. Kulshan glaciomarine stony clay was exposed beneath the Deming sand a few hundred yards to south along Highway 9 where marine fossils were abundant. A radiocarbon date of 11,970 ± 280 radiocarbon years was obtained from these shells. The Bellingham glaciomarine stony clay, Deming sand, and Kulshan glaciomarine stony clay are all present here, just as at the type locality at the end of the Smith road and at Bellingham Bay.
		Turn south on Highway 9 and drive a few hundred yards to the Kulshan fossil locality in the roadcut on the left (east) side of the road. The most abundant marine fossils in the Kulshan glaciomarine stony clay are just above Chuckanut sandstone, which outcrops at the south end of the roadcut, but are usually obscured by vegetation and may require digging out. Return to Mt. Baker Highway.
		View of the Van Zandt landslide from various points near Deming.
19.6	2.3	Welcome, junction with the Mosquito Lake road to the Middle Fork valley. Travel across a flat outwash terrace. Wood in glaciomarine stony clay in a gully to the right of the road down to the Middle Fork bridge has been dated at 11,910 ± 80 radiocarbon years. This is very close to the age of the Kulshan glaciomarine stony clay at the type locality near Cedarville, at Deming, and at Bellingham Bay.
22.0	2.4	Huge boulders of the Racehorse Creek landslide occur to the right of the highway but are now largely obscured by vegetation.
23.2	1.2	Coal Creek. Ascend a late Pleistocene outwash terrace at the lower end of Columbia Valley, an abandoned outwash outlet from ice at the north end of the valley at Cultus Lake.
		Kendall Creek. View ahead to the left is to American Sumas Mt. underlain by rocks of the Chilliwack Group. The Chilliwack here consists of Paleozoic sedimentary and volcanic rocks. View to the right ahead is Red Mt., also made up of Chilliwack sedimentary and volcanic rocks, including late Paleozoic limestone. To the right across the Nooksack River is Slide Mt. composed of Chuckanut sandstone. A large landslide has occurred on the north slope of the mountain.
25.4	5.8	Kendall. Junction with Highway 547 to Columbia Valley. To the left, two landslides extend as lobes into the valley.
26.2	0.8	Kendall moraine. The highway bends around the ridge making the moraine and a deep roadcut exposes poorly sorted glacial till containing many glacially faceted and striated pebbles and cobbles. The Kendall moraine is composed of ~40 ft. of till, containing abundant glacially faceted and striated boulders and cobbles derived from the upper North Fork valley. The number of cobbles and boulders in the till that are well faceted and striated is striking—almost all of them show evidence of glacial transport, probably because the valley glacier was so long. The lithology of cobble

		and boulders in the Kendall moraine is dominated by Chuckanut sandstone (54%), Mt. Baker andesite (12%), and greenstone/sandstone composed of volcanic grains (14%) (presumably from the Nooksack Group). The composition of these rocks demonstrates that the main source of ice was from the North Fork to the east, rather than from the Cordilleran Ice Sheet to the north. The Mt. Baker andesite was probably contributed from Mt. Baker via Bagley Creek, Glacier Creek, and Wells Creek. Chuckanut sandstone was derived from the valley sides immediately upvalley, and the graywacke/ greenstone came from the valley sides near Glacier.
28.1	1.9	Drive across a low moraine, enter Maple Falls. Hills to the north are composed of Paleozoic Chilliwack greenstones, volcanic sandstone, and limestone. Maple Falls is built on a late glacial outwash terrace which received most of its sand and gravel from a glacier margin at Silver Lake to the north. Leaving Maple Falls drop off the terrace and onto the modern Nooksack floodplain.
30.5	2.4	Boulder Creek. For many years, Boulder Creek shed large volumes of cobbles, boulders, and trees during storms, and road closures were not uncommon. The problem is unstable slopes in the upper drainage that resulted in slope failures. A substantially higher bridge has recently replaced the old bridge.
31.5	1.0	Roadcut in steeply dipping, thin-bedded Chuckanut sandstone and shale containing 50–million–year–old palm fronds and other plant fossils. The Chuckanut sandstone is separated from the Chilliwack Group rocks to the north by the Boulder Creek fault.
31.7	0.2	Parking area on right (formerly a view point of Mt. Baker but now obscured by vegetation). About 100 yards back down the highway, Chuckanut sandstone and black shale contains leaf fossils.
32.0	0.3	View of a large landslide on Slide Mt. across the Nooksack River.
33.0	1.0	Cross the Nooksack Rive
33.9	0.9	View ahead of Skyline Divide underlain by sandstone and shale of the Nooksack Group. Mt. Baker to the right.
		Travel across irregular topography with many mounds and depressions on the surface of the huge Church Mt. landslide that occurred 2,500 years ago. The Church Mt. landslide fills the Nooksack valley for 7 miles and is more than 300 feet thick. It extends through the town of Glacier all the way to Douglas fir campground.
34.6	0.7	Cross Cornell Creek.
35.3	0.7	Town of Glacier. View of Church Mt. composed of volcanic and sedimentary rocks of the upper Paleozoic Chilliwack Group, which overlie volcanic sandstone and shale of the Jurassic Nooksack Group. The older Chilliwack Group rocks have been pushed over the younger Nooksack Group along the Church Mt. thrust fault.
35.6	0.3	Cross Glacier Creek. The muddy water is due to large amounts of glacial silt discharged by the Coleman and Roosevelt glaciers at its headwaters.

35.7	0.1	Glacier Ranger Station, US Forest Service. Stop and take a look at the cross section of a huge, 800–year–old tree on display. The visitor center has maps and information and usually has some interesting rocks on display, including 175–million–year–old fossil marine shells from the Nooksack Group.
36.4	0.7	Junction with Glacier Creek road on the right. The road extends up Glacier Creek to the trailhead to the Coleman glacier. Excellent views of Mt. Baker and the Coleman and Roosevelt glaciers. The straight stretch of road ahead is across the upper part of the Church Mt. landslide, which is here several hundred feet thick. Some of the landslide debris has rocks showing signs of ore mineralization that led to a mine driven into the slide debris on the right side. Good views of Church Mt. straight ahead
37.4	1.0	Cross the Nooksack River, Douglas fir campground on left.
37.6	0.2	Junction with Canyon Creek road on the left.
37.8	0.2	Cross Coal Creek.

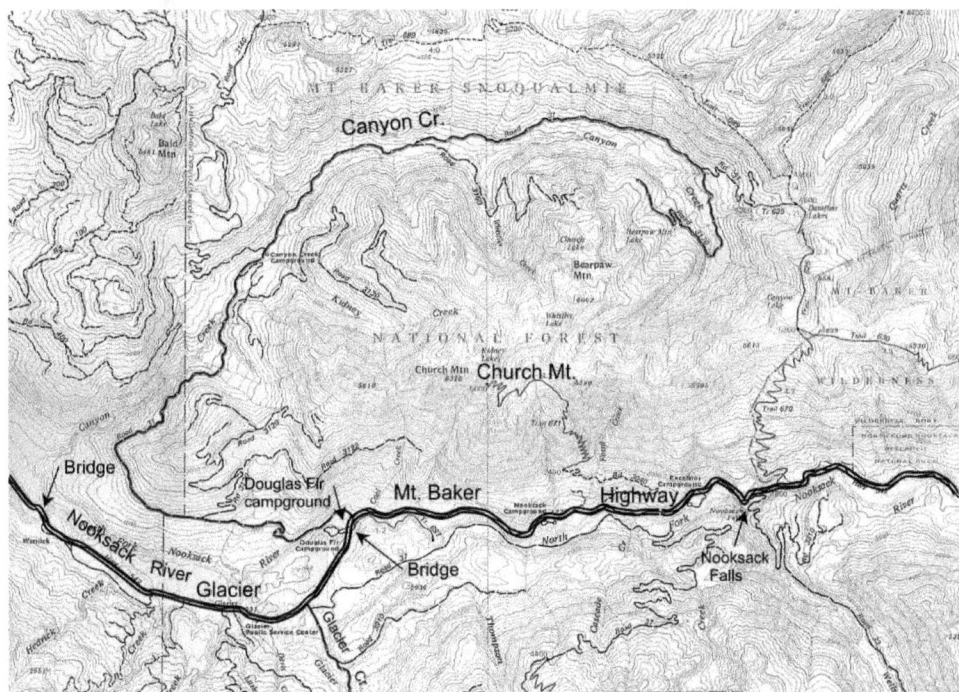

Figure 304. Map of field trip route from Nooksack River bridge to Nooksack Falls. (USGS topographic map)

317

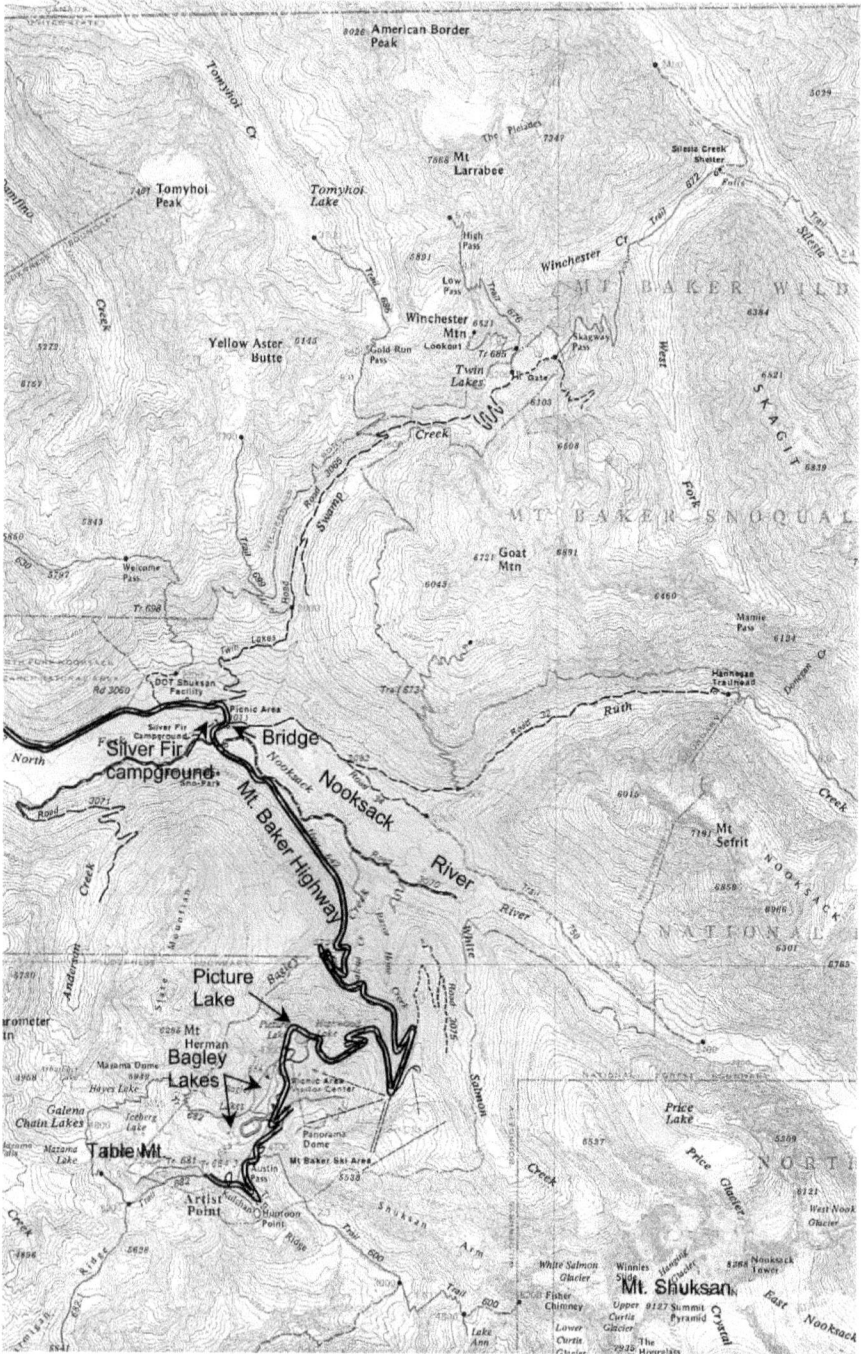

Figure 305. Map of field trip route from Silver fir campground to Artist Point. (USGS topographic map)

38.2	0.4	Old quarry on the left (poorly visible from the highway) exposes upper Nooksack Group green sandstone made up of sand derived from volcanic rocks and some dark shale, slightly altered by heat and pressure. 175–million–year–old fossil oysters (*Buchia*) and and some small ammonites have been found here. The massive sandstone bed is resistant to erosion and makes a topographically conspicuous ridge that rises eastward on the north side of the valley as part of the west flank of a large anticline.
	0.9 to 1.4	Roadcuts on left in lower Nooksack Group dark colored volcanic-derived sandstone and shale.
40.4	0.8	Dark gray, massive, volcanic-derived sandstone and siltstone containing sparse marine fossils.
40.6	0.2	Roadcuts in dense, black Nooksack Group siltstone.
41.6	1.0	Fossil Creek. Cobbles and boulders of Nooksack Group siltstone in the creek bed contain 175–million–year–old fossil oysters (*Buchia*) and some belemites and, rarely, ammonites.
41.7	0.1	Road cut in upper Jurassic Nooksack siltstone on the left.
41.8	0.1	Site of old Deadhorse Creek bridge that formerly provided automobile access to Deadhorse Creek, then later converted to a footbridge.
41.9	0.1	Bridge Forest Camp.
42.0	0.1	Roadcut on the left in basal Nooksack green, volcanic-derived sandstone, conglomerate, and siltstone.
42.4	0.4	Wells Creek volcanic breccia beneath Nooksack sedimentary rocks in the core of the Wells Creek anticline.
42.7	0.3	Wells Creek volcanic breccia.
43.2	0.5	Wells Creek volcanic greenstone (lava altered by low heat and pressure) and black slate.
43.3	0.1	Excelsior Forest Camp
	0.3 to 0.5	Roadcuts in volcanic greenstones on the left
44.0	0.2	Roadcuts in Wells Creek volcanic greenstone, breccia and slate in the core of the Wells Creek anticline. Junction with the road to Nooksack falls and Wells Creek. Turn right and drive to parking area at the top of the falls. Here the Nooksack River plunges over a cliff of resistant volcanic rocks to form Nooksack falls, which are due to the difference in resistance to erosion between the hard volcanic rocks and the relatively less resistant marine sedimentary rocks of the overlying Nooksack Group. Return to Mt. Baker Highway.
44.4	0.4	Massive outcrops of Wells Creek volcanic greenstone (lava altered by low heat and pressure) with some interbedded volcanic breccia. Observation turnouts on the right overlook a deep canyon of the Nooksack River.
44.7	0.3	Observation point on the right. Volcanic greenstones.
44.9	0.2	Dense siltstone and shale on the left.

45.0	0.1	Foliated volcanic breccia and slate overlain to the east by nearly vertical altered lava.
45.2	0.2	A cliff of Mt. Baker lava with prominent columnar jointing is visible across the Nooksack River to the south. This ridge-capping lava is a remnant of lava that flowed down an ancient valley from Mt. Baker but now stands as a mesa about half a mile long about 700 feet above the floor of the modern Nooksack River. The lava flow is about 300 feet thick, with prominent columnar jointing (fractures) that formed during cooling and contraction of the lava. Broken columns now litter the slope at the base of the cliff. The ridge-capping topographic position suggest and early isotope dating of 400,000 ± 200,000 years suggested correlation with similar dates from Table Mt. lava flows, but more recent isotope dating of the lava gave an age of 202,000 ± 9,000 years, about 100,000 years younger than recent dates of 300,000 years for the Table Mt. flows.
	0.1 to 0.3	The volcanic greenstone is overlain by dark slate and breccia folded into small scale anticlines and synclines in the lower Wells Creek volcanic sequence. A few belemites occur in the slate. At the easternmost exposure, slate dips steeply eastward and is cut by subhorizontal fracture fillings.
45.7	0.2	Cliff of massive volcanic greenstone on the left.
46.0	0.3	Sheared volcanic greenstone.
46.2	0.2	Foliated volcanic breccia.
48.3	2.1	The cliff north of the road consists of Nooksack black slate.
48.8	0.5	Roadcuts in Nooksack thin–bedded black slate and siltstone.
49.0	0.2	Nooksack black shale, siltstone, and green sandstone derived from volcanic sources.
49.2	0.2	200–foot long roadcut in Wells Creek volcanic greenstone and volcanic breccia on the east limb of a syncline.
49.4	0.2	Road to Welcome Pass trail.
49.8	0.4	Shuksan highway station. Twin Lakes road on left.
50.1	0.3	Outcrop of Chilliwack volcanic greenstone on left.
50.2	0.1	Cross the Nooksack river. Silver Fir Campground. Outcrops of Chilliwack volcanic greenstone on left.
	0.7 to 2.3	Semi-continuous roadcuts in Chilliwack volcanic greenstone and volcanic breccia.
54.8	2.3	Mt. Baker lava flow with columnar jointing. Larger white crystals are plagioclase feldspar. Cross Bagley Creek. Compact sandy clay glacial till.
	0.5 to 1.0	Chilliwack volcanic greenstone and volcanic breccia is exposed east of Razorhone Creek. The Shuksan thrust fault occurs in an interval directly east covered by glacial deposits. On the east, a roadcut exposes Darrington phyllite, showing intense deformation related to the thrust faulting. The rocks are intensely ground up in the fault zone and quartz veins are broken and stretched into lens–shaped forms. Many minor folds occur, along with minor, spoon–shaped thrust faults.

56.5	0.7	From last outcrop, discontinuous exposures of Darrington phyllite.
56.7	0.2	Cross Razorhone Creek. Chilliwack volcanic greenstone west of the creek.
57.5	0.8	Mt. Baker lava flow resting on Chilliwack volcanic greenstone.
57.8	0.3	Mt. Baker lava flow on Chilliwack volcanic greenstone.
57.9	0.1	Chilliwack volcanic greenstone
58.6	0.7	Mt. Baker lava flow with columnar jointing. exposed for about half a mile.
60.4	1.8	Scenic and geologic view of Mt. Shuksan. Distant view of Red Mt. (Mt. Larabee), Mt. Tomyhoi, Yellow Astor Butte.
60.6	0.2	Heather Meadows picnic area.
61.0	0.4	Chilliwack volcanic greenstone
62.9	1.9	Mt. Baker lava flow with horizontal fractures that break the rocks into parallel slabs about one inch thick.
63.0	0.1	End of road, parking area. Spectacular views of Mt. Shuksan, Mt. Baker, and peaks to the north.

INDEX

www.ingramcontent.com/pod-product-compliance
Lightning Source LLC
Chambersburg PA
CBHW081239220326
41597CB00023BA/4130